T0321670

COMPREHENSIVE UTILIZATION OF CITRUS BY-PRODUCTS

COMPREHENSIVE UTILIZATION OF CITRUS BY-PRODUCTS

YANG SHAN

Editor-in-Chief

Amsterdam • Boston • Heidelberg • London
New York • Oxford • Paris • San Diego
San Francisco • Singapore • Sydney • Tokyo
Academic Press is an imprint of Elsevier

Academic Press is an imprint of Elsevier
125 London Wall, London EC2Y 5AS, United Kingdom
525 B Street, Suite 1800, San Diego, CA 92101-4495, United States
50 Hampshire Street, 5th Floor, Cambridge, MA 02139, United States
The Boulevard, Langford Lane, Kidlington, Oxford OX5 1GB, United Kingdom

Notices
Knowledge and best practice in this field are constantly changing. As new research and experience broaden our understanding, changes in research methods, professional practices, or medical treatment may become necessary.

Practitioners and researchers must always rely on their own experience and knowledge in evaluating and using any information, methods, compounds, or experiments described herein. In using such information or methods they should be mindful of their own safety and the safety of others, including parties for whom they have a professional responsibility.

To the fullest extent of the law, neither the Publisher nor the authors, contributors, or editors, assume any liability for any injury and/or damage to persons or property as a matter of products liability, negligence or otherwise, or from any use or operation of any methods, products, instructions, or ideas contained in the material herein.

Library of Congress Cataloging-in-Publication Data
A catalog record for this book is available from the Library of Congress

British Library Cataloguing-in-Publication Data
A catalogue record for this book is available from the British Library

ISBN: 978-0-12-809785-4

For information on all Academic Press publications visit our website at https://www.elsevier.com/

Publisher: Jonathan Simpson
Acquisition Editor: Simon Tian
Editorial Project Manager: Vivi Li
Production Project Manager: Susan Li
Designer: Matthew Limbert

Typeset by TNQ Books and Journals

CONTENTS

Preface vii

1. Functional Components of Citrus Peel 1

1. Essential Oils 1
2. Monoterpenes 2
3. Pectin 3
4. Flavonoids 4
5. Limonoids 8
6. Carotenoids 8
7. Cellulose 10
8. Synephrine 11
9. Other Functional Components 12

2. Methods for Determining the Functional Components of Citrus Peel 15

1. Chemical Analysis: Determination of Pectin Using a Gravimetric Method 16
2. Spectral Analysis: Determination of Total Flavonoids in Citrus Peel by $AlCl_3$-UV-Visible Spectroscopy 18
3. Determination of Multi-Methoxyflavones in Citrus Peel by High-Performance Liquid Chromatography (NY/T 2336-2013) 20
4. Determination of Aromatic Components in Citrus Peel by GC–MS 24
5. LC–MS (HPLC–MS) 29
6. HPLC-Nuclear Magnetic Resonance Spectroscopy 29

3. Extraction Processes of Functional Components From Citrus Peel 31

1. Classification and Extraction Processes of Essential Oils From Citrus Peel 33
2. Extraction Process of Pectin From Citrus Peel 41
3. Extraction Processes of Flavonoids From Citrus Peel 49
4. Extraction of Limonoid and its Analogs 51
5. Extraction Process of Synephrine From Citrus Peel 53
6. Extraction Process of Edible Cellulose Powder 54
7. Extraction Process of Seed Oil 56

4. Isolation and Structural Identification of Flavonoids From Citrus 59
1. Technological Process for the Separation and Purification of Flavonoids From Citrus Peel 60
2. Structural Identification of Flavonoids From Citrus Peel 62

5. Five Types of Semisynthetic Bioactive Flavonoids From Hesperidin 65
1. Synthesis Route of Bioactive Flavonoids 66
2. Synthesis of Bioactive Flavonoid Compounds 66
3. Key Points in Flavonoid Synthesis 71

6. Drying of Citrus Peel and Processing of Foods and Feeds 75
1. Drying of Citrus Peel 76
2. Processed Foods of Citrus Peel 78
3. Production of Various Citrus Feed Additives 80

7. Biotransformation of Citrus Peel 85
1. Production of Food 85
2. Production of High-Protein Feeds 86
3. Production of Ethanol, Mushroom Cultivation, and Preparation of Other Biological Products 87

8. Production of Biodegradable Packages Using Citrus Peel 91
1. Technological Process 91
2. Key Operating Points 91

9. Processing Equipment for Citrus Peel By-products 93
1. Drying Machines and Equipment 93
2. Recycling Equipment for Aromatic Substances in Citrus Juice 101
3. Typical Citrus Processing and Production Line 104

Appendix *107*
References *109*
Index *113*

PREFACE

China is the major source of origin for citrus in the world and the leading citrus-producing country. In 2011 the total plantation area of citrus in China was 2.211 million hectares, with an annual production of up to 27.5 million tons, both of which were the highest in the world. Citrus peel is a by-product of the citrus-processing industry, which accounts for 40–60% of the total weight of citrus. At present, treatment of citrus peel is the biggest problem in the citrus-processing industry in China. Pomace weighing more than 10 million tons is produced annually in China. This pomace contains 130,000 tons of pectin, 30,000 tons of essential oils, and 10,000 tons of flavonoids, and has a total value of up to 350–400 billion yuan. However, only a small amount of sun-dried tangerine peel is utilized, and a large amount of citrus peel is buried and burned. These treatments pollute the environment and waste resources.

The shortage and depletion of global resources is an indisputable fact. The comprehensive utilization of resources by the agro-product-processing industry is an important research area both domestically and overseas. Research and development of technology that can facilitate conservation and the efficient transformation of citrus by-product resources could improve resource consumption and save resources. These approaches could also resolve environmental pollution problems, improve the ecological environment, and enhance resource utilization during citrus processing, thereby facilitating the sustainable development of the citrus industry.

On the basis of domestic and foreign literature and documents, the authors collated long-term research achievements and experience in this field and wrote this book, *Comprehensive Utilization of Citrus By-products*. The aim of the book is to provide a reference for citrus by-product utilization by scientists, scholars, and students from advanced institutes, universities, and citrus-processing enterprises, as well as farmers from cooperative organizations related to citrus processing.

This book was written by Yang Shan as the chief editor, with Gaoyang Li, Jianxin He, Donglin Su, Juhua Zhang, Fuhua Fu, Qun Zhang, Xiangrong Zhu, Wei Liu, Lvhong Huang, Jiajing Guo, Qiutao Xie, and other colleagues participating in the compilation. It comprises nine chapters in the main text, with one chapter in the appendix, including standard lists of citrus by-products from home and abroad, and major references. The main text considers each link in the overall citrus by-product-processing industry chain, including the functional components of citrus peel; determination methods and extraction processes for functional components; isolation and structural identification of flavonoids; synthesis of flavonoid compounds; production of biodegradable packaging materials, such as scattered-transplantation seedling plates using citrus peel; citrus by-product-processing equipment; and related standards at home and abroad. This book features complete content, as well as both scientific and practical guidance.

Some mistakes and inadequacies are inevitable because of limitations in the knowledge and expertise of the authors. Criticism from experts and readers would be highly appreciated.

Yang Shan

February, 2016

CHAPTER 1

Functional Components of Citrus Peel

Contents

1. Essential Oils 1
2. Monoterpenes 2
3. Pectin 3
4. Flavonoids 4
5. Limonoids 8
6. Carotenoids 8
7. Cellulose 10
8. Synephrine 11
9. Other Functional Components 12

During citrus processing, a large amount of citrus peel is produced, which comprises approximately 25% of the total weight of citrus fruit. Citrus peel contains several functional components, such as essential oil, pectin, carotenoids, hesperidin, and limonin, which are important raw materials in the chemical and pharmaceutical industries. The utilization of functional components of citrus peel has become an important part of the citrus-processing industry. The following discussion describes the types, applications, and utilization status of the functional components of citrus peel.

1. ESSENTIAL OILS

An essential oil is a concentrated hydrophobic liquid that evaporates at room temperature. It contains volatile aromatic compounds from plants. Oil cells in citrus peel are rich in essential oil, which comprises approximately 0.5–2% of the fresh weight of the peel. Essential oil can be used as a flavoring and perfuming agent in alcohol and tobacco products, beverages, condiments, candies, and pastries. It is also used extensively in the production of daily consumables, essences, and pesticides. The essential oil from citrus peel also has a sedative effect on the human central nervous system, which can mitigate stress and remove fatigue. Moreover, despite its trace levels in essential oil, coumarin has an obvious anticancer effect; it

Comprehensive Utilization of Citrus By-products
ISBN 978-0-12-809785-4
http://dx.doi.org/10.1016/B978-0-12-809785-4.00001-0

probably acts by decomposing the toxic functional groups in carcinogenic substances and inhibiting the metabolism of cancer cells. Previous studies have demonstrated that the essential oil from citrus peel also plays roles as an expectorant, cough inhibitor, gastrointestinal motility promoter, and digestive juice secretion enhancer. It also alleviates pain, dissolves gallstones, and relieves inflammation. In addition, the essential oil from citrus peel has obvious efficacy in human beauty and health care.

According to statistical analyses, essential oils are added to products that account for sales of $500 billion throughout the world, where the value of the essential oils is estimated to be $16–20 billion. The annual global production of citrus essential oil is approximately 16,000 tons, and its cost on the international market is $14,000/ton. Therefore, citrus essential oil is in great demand in the international market and has promising market prospects.

2. MONOTERPENES

Monoterpenes are widely distributed in various citrus peels, especially in the essential oil from citrus peel. d-limonene and d-carvone are the two most important types of monocyclic terpene compounds in citrus (Fig. 1.1). d-limonene is a specific major aromatic component of citrus. Citrus fruits (especially citrus peel) contain a high essential oil content. In lemon and orange peels, the mass fraction of the limonene content of essential oil is up to 90–95% (w/w).

Limonene is an effective natural solvent for removing grease and dirt. In most cases, limonene can be applied in home and mechanical devices as a substitute for corrosive alkaline detergents. In addition, it can be used as an important component of strong delousing agents. Previous studies have confirmed its function in inhibiting the excessive excitation of the central nervous system. Recent studies have also demonstrated that limonene has a role in reducing stress. Because limonene can inhibit the activity of HMG-CoA reductase as a rate-limiting enzyme, it is possible that it inhibits cholesterol synthesis. Another study found that the administration of a

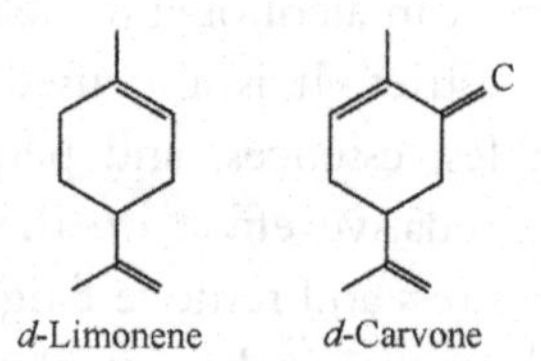

Figure 1.1 Major monoterpene compounds in citrus.

mixture containing 97% d-limonene via biliary drainage could effectively dissolve the remaining cholesterol gallstones in patients subjected to surgery. Moreover, the anticancer effect of limonene is attracting increasing attention. An increasing number of studies have shown that monoterpenes from citrus have an obvious inhibitory effect on tumors caused by chemical carcinogens. Increasing evidence also supports the role of limonene in cancer prevention and treatment, which has been validated by its chemopreventive roles during the initiation or development stage in various in vivo tumor systems, such as rodent models of breast, skin, liver, and lung cancers, as well as a precancerous model induced by chemical carcinogens.

3. PECTIN

The yellow skin, white skin, flesh, juice sac, and central column of citrus fruits contain an abundance of pectin, which exists in the form of propectin. The pectin contents of different tissues and varieties differ considerably. In decreasing order, the pectin content and quality are as follows: lime > lemon > grapefruit > orange > loose-skin citrus. The pectin content is generally 20–30% of the dry weight of citrus peel. Pectin is a thickener and stabilizer, which is widely used in the food industry. Pectin is a polysaccharide with 1,4-galacturonic acid as the basic structure, whose structure is shown in Fig. 1.2. The percentage of the esterified carboxyl group in pectin is defined as the esterification degree. Pectin with an esterification degree of more than 50% (methoxyl group content more than 7%) is called high-ester pectin. In contrast, pectin with an esterification degree of less than 50% (methoxyl group content less than 7%) is called low-ester pectin.

Pectin is a good thickener and stabilizer and is widely applied in the food industry. The dietary fiber in pectin is an important material for maintaining a healthy body. It has roles in enhancing gastric motility and nutrient absorption. Furthermore, the dietary fiber in pectin can have

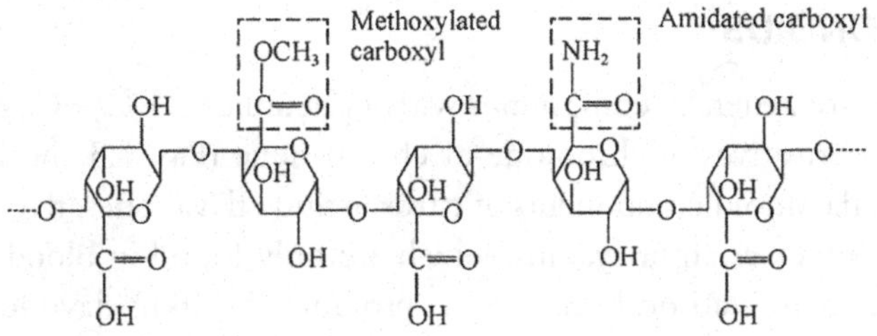

Figure 1.2 Structure of pectin.

preventative and therapeutic effects on high blood pressure, cancer, diabetes, obesity, and other diseases. Because of its satiating effect, pectin can be used as a weight loss agent. The addition of pectin to jam and juice not only stabilizes jam during transportation but also improves the flavor and reduces syneresis thereby producing an irreplaceably excellent taste. Pectin also has the characteristic of acid resistance when added to fruit juice and syrup, which can stabilize suspended oily lactic acid, thereby resulting in juice with a fresh flavor. The addition of pectin to dairy products can play an important role in preservation and stability. It prevents the coagulation of casein protein and extends the shelf life of acidic dairy products. Moreover, pectin can be used by the pharmaceutical industry to produce laxative and hemostatic agents or plasma substitutes. Importantly, pectin is also a preventative agent or antidote used to treat exposure to heavy metal ions.

Currently 70% of the commercial pectin used throughout the world is extracted from citrus peel. Thus, citrus peel is a highly valuable raw material for the extraction of pectin. Pectin has excellent domestic and overseas market prospects. At the end of the 20th century, the annual worldwide trade volume of pectin was approximately 30,000 tons, which accounted for 10% of the total food gums (300,000 tons), with an annual growth rate of 4–5%. At present, the price of pectin is approximately RMB 100,000–130,000 yuan/ton. Annual pectin production in China is only 2000–3000 tons, and the quality and production amount is inferior to that produced in Copenhagen, Denmark. Thus, a large amount of pectin still needs to be imported. Overall, pectin is still in the initial stage as a food additive in China. Research and development into pectin production has been carried out, but the quality and quantity of pectin produced is still not as good as required. The developing food and food additive industries require large-scale production of high-quality pectin using a large number of renewable resources, such as citrus peel and apple pomace.

4. FLAVONOIDS

Flavonoids are generally compounds with a phenolic C_6–C_3–C_6 structure, which are derivatives of chromone or chromanene (Fig. 1.3 and Fig. 1.4). The major flavonoid components of citrus include flavanones, flavones, and flavonols, as well as anthocyanins, which are only found in blood oranges.

According to statistical analyses, approximately 4000 flavonoid compounds have been isolated and identified. These compounds exist as free

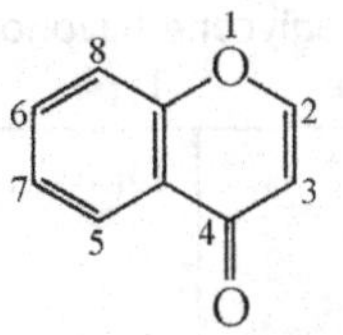

Figure 1.3 Structure of chromone (benzopyrazole).

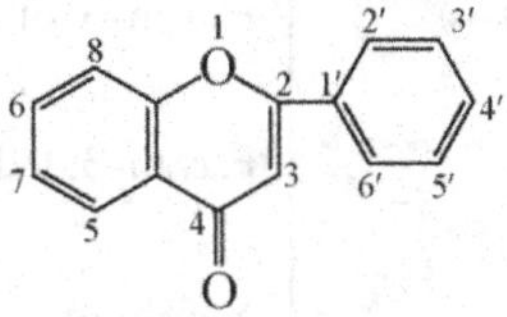

Figure 1.4 Structure of 2-phenyl chromone.

forms of aglycones or glycosides, which are widely distributed in the plant kingdom. On the basis of the presence or absence of a double bond between the C_2 and C_3 atoms in the B ring, the hydroxyl group at the C_3 position, the connection site between the B ring and C ring, and the open or closed status of the C ring in the molecular structure, natural flavonoids can be classified as flavones, flavanones, flavonols, anthocyanidins, dihydroflavonols, isoflavones, dihydroflavonoids, isoflavonols, chalcones, dihydrochalcones, dibenzene pyridine ketones, and orange ketones. Among these compounds, lemons only contain some flavonols but anthocyanins mainly exist in red oranges. Depending on the oxidation degree of the 3-carbon chain, the connection position of the B ring (phenyl group) at position 2 or 3, and the ring structural characteristics between the 3-carbon chain and B ring, the major natural flavonoids can be classified into the structures shown in Table 1.1.

Flavanone is one of the most abundant flavonoid compounds in citrus, hesperidin is a glycoside and the major flavonoid compound in orange. Hesperidin, naringenin, and rutin are the major flavonoid compounds in pomelo peel, where the content is up to 80%. These compounds are found mainly in the peel and pulp of pomelo, grapefruit, and bitter orange and are the major bitter substances in these fruits. At present, over 60 flavonoid compounds have been isolated and identified from citrus, the most common of which are naringin, hesperidin, neohesperidin, and narirutin. The structures of several major flavonoids from citrus are shown in Fig. 1.5. In citrus-processing waste, flavonoids mainly come from concentrated solutions of fruit peel and parts of fruit tissues, which are represented by hesperidin in oranges and naringin in grapefruits.

Table 1.1 Structures and types of aglycone flavonoids

Type	Basic structure	Type	Basic structure
Flavonoids	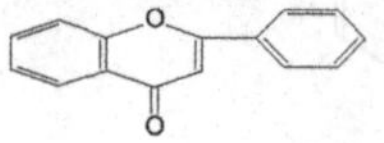	Dihydrochalcones	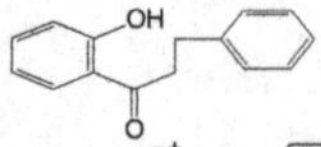
Flavonols	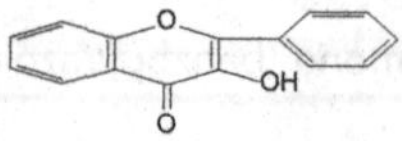	Anthocyanins	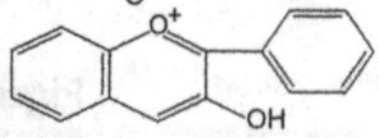
Flavanone	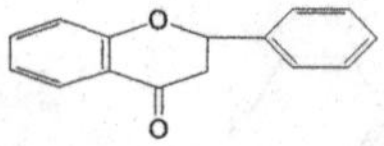	Flavan-3-ol	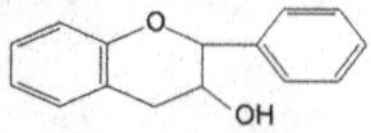
Dihydroflavonol	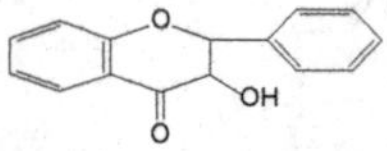	Flavan-3,4-diol	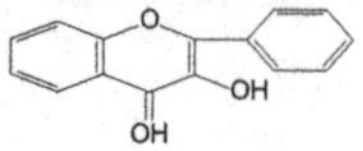
Isoflavones	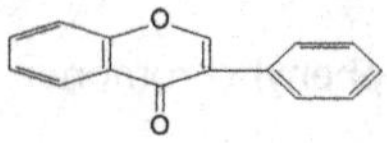	Xanthone	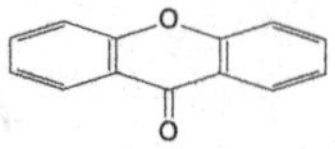
Isoflavanone	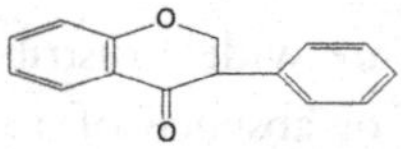	Aurone	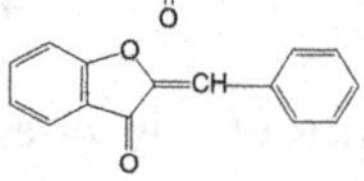
Chalcone	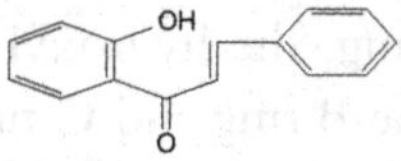		

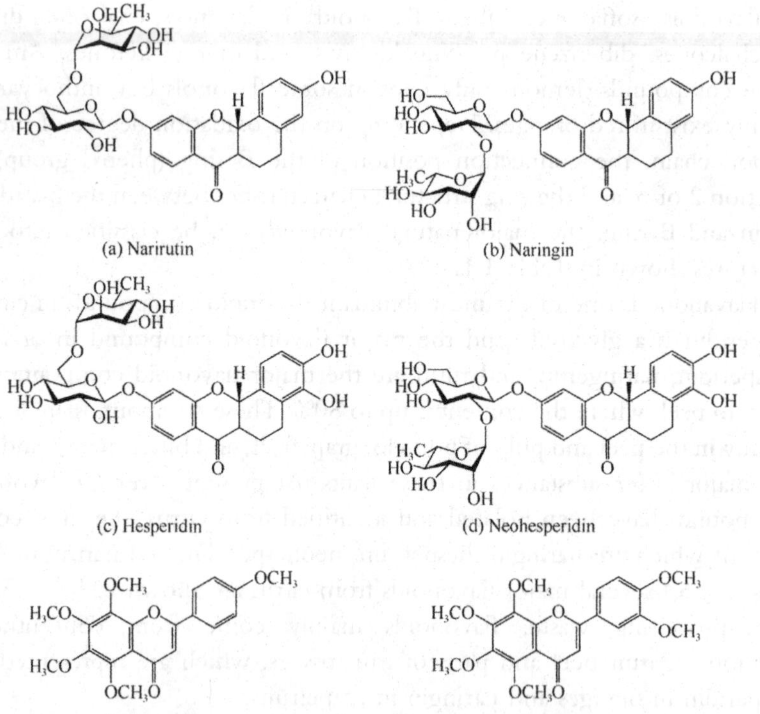

(a) Narirutin

(b) Naringin

(c) Hesperidin

(d) Neohesperidin

(e) Tangertin

(f) Nobiletin

Figure 1.5 Structures of flavonoids in citrus.

Hesperidin is a flavanone glycoside, which has an important role in maintaining the normal osmotic pressure of blood vessels, reducing vascular fragility, increasing capillary toughness, and shortening bleeding time. Some data indicate that hesperidin can reduce cholesterol in the body and has antivirus and immune-promoting functions. Hesperidin can be converted to dihydrochalcone via hydrogenation, which is a high-intensity (approximately 300 times sweeter than sucrose) low-calorie sweetener that is suitable for diabetic and obese patients. Hesperidin can reduce cholesterol in the body. It can be used as an adjuvant therapy for cardiovascular diseases in clinical applications. It can be prepared as drugs for preventing atherosclerosis and myocardial infarction. Liang et al. found that under certain concentrations and pH conditions, hesperidin has antioxidant and antifungal effects in the presence of synergistic interactions with some drugs. Wilson et al. reported that hesperidin can improve the symptoms of guinea pigs with vitamin C deficiency. It increases vitamin C levels in the adrenal glands and spleens of guinea pigs. Wacker found that hesperidin can prevent virus infections. A vitamin P-like function was demonstrated for hesperidin, but with no side effects or irritation of the skin. Therefore, hesperidin is an important raw material for the food, medicine, and cosmetics industries.

Naringin, ie, namenaringenin-7-*O*-neohesperidoside, can be used as a natural pigment, flavor enhancer, and bittering agent during the production of foods and beverages in the food industry. It is also a raw material for the synthesis of high-intensity, nontoxic, and low-energy new sweeteners such as dihydrochalcone and neohesperidin dihydrochalcone. In the pharmaceutical industry, naringin is used in the production of drugs for the prevention of cardiovascular diseases and anti-inflammatories. Naringin can also be used to produce a range of high value-added organic materials, such as rhamnose, limonin, acidic azo dyes, and other plant colorings. Nogata et al. reported the bioactive antioxidant capacity and anticancer activity of naringin. Many studies have shown that naringin also functions in preventing poisoning, alleviating spasms and pain, reducing capillary permeability, decreasing the incidence of osteoarticular diseases, and enhancing the absorption and metabolism of a number of drugs in the human body.

Neohesperidin is the precursor of a sweet substance found in citrus peel. The natural extract obtained from citrus peel, which is also named neohesperidin, can be hydrogenated in an autoclave to produce a new sweetener called neohesperidin dihydrochalcone, which has sweetness equivalent to 1500–1800 times that of sucrose. Neohesperidin dihydrochalcone has been widely used in foods, drinks, and toiletries such as toothpaste as a sweetener to replace sodium saccharin in Europe. It is also one of the best medicinal sweeteners and flavoring agents that has been discovered to date. Researchers from Western countries recently reported that neohesperidin dihydrochalcone can inhibit the secretion of gastric acid, which means that it can be used as an antacid agent in combinatorial applications with aluminum hydroxide gel or other antacid agents to enhance the antistomach acid effects of the latter. Therefore, neohesperidin dihydrochalcone can be used as a raw material for the production of novel drugs for stomach diseases.

Flavonoids have become essential ingredients in the human diet because of their widespread presence in plants. According to statistical analyses, the per capita daily intake of flavonoids is up to 1 g in the United States.

5. LIMONOIDS

Limonoids are a class of triterpenoids with a furan ring, which exist as free aglycones or glycosides that are identical to flavonoid compounds. Limonoids are distributed mainly in Rutaceae fruits, such as *Citrus aurantium* L., *Citrus sinensis* Osbeck, *Citrus medica* L., and grapefruit. Higher levels of limonoids have been detected in the kernels (seeds), with approximately 0.05–0.1‰ in the peel. At present, 36 types of limonoids and 17 types of limonoid glycosides have been separated from citrus. The levels of limonin and nomilin, which are representative compounds with strong bitterness, can reach up to 6 mg/kg in the juice, but these high levels affect the flavor. The chemical structures of several major citrus limonoids are shown in Fig. 1.6.

Limonoid glycosides exhibit not only excellent water solubility and low bitterness but also physiological activities similar to those of aglycones, such as antitumor activity. Previous studies have shown that limonoid glycosides have good developmental prospects and could be used as basic raw materials for functional foods. Anticancer and insect-repellent capacities are the major biological activities of limonoids. In addition, some limonoids have antibacterial and antiviral functions.

6. CAROTENOIDS

Citrus peel contains two types of natural pigments with different characteristics: lipid-soluble carotenoids and water-soluble yellow pigments. The representative compounds are a water-soluble pigment (orange pigment A)

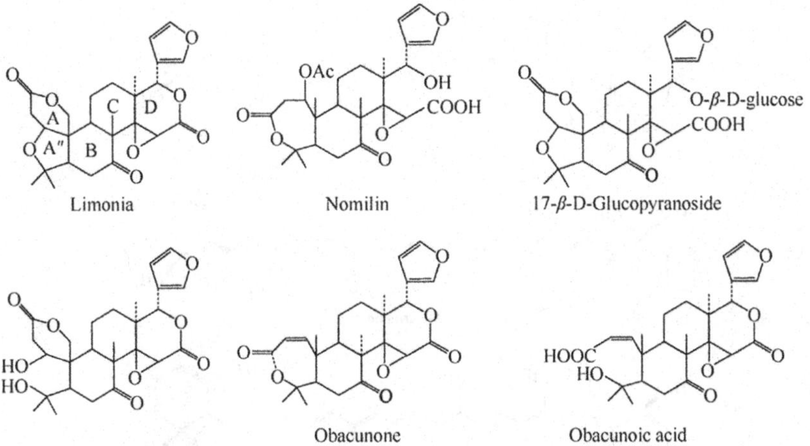

Limonia

Nomilin

17-β-D-Glucopyranoside

Obacunone

Obacunoic acid

Figure 1.6 Chemical structures of limonoids in citrus.

and a lipid-soluble pigment (orange pigment B), both of which have high safety and excellent nutritional and medicinal values. The pigment derived from citrus peel is an important natural pigment, the major component of which is a mixture of carotenoids that can replace synthetic pigments in food colorings. Carotenoids comprise various lipid-soluble plant pigments, which contain forty carbon atoms and eight isoprenoid units. Carotenoid contents and compositions are determinants of the color of citrus peel. The major carotenoids in citrus are α-carotene, β-carotene, lutein, zeaxanthin, and β-cryptoxanthin, and their structures are shown in Fig. 1.7.

Carotenoids have two health care effects: (1) an antioxidative activity that allows carotenoids to protect cells from oxidative damage and reduce the incidence of diseases, such as arteriosclerosis, cancer, and arthritis; (2) α-carotene and β-carotene are vitamin precursors, which can be transformed into vitamin A in the intestinal mucous membrane. β-carotene can effectively prevent or inhibit the occurrence of diseases, slow down the aging process, and promote immune functions, with an especially important physiological role in cancer prevention.

Carotenoids comprise approximately 0.1–0.5% of the dry weight of citrus peel. Carotenoids are safe colorants. They also contain vitamin E and the rare element selenium (Se); therefore, they may prevent the growth of cancer cells and delay the senescence of cells, as well as enhance human immunity. In addition to their roles as coloring agents, they are important ingredients in food supplements, flavoring agents, and health foods. The current annual global demand for β-carotene is 1200–1500 tons, and its

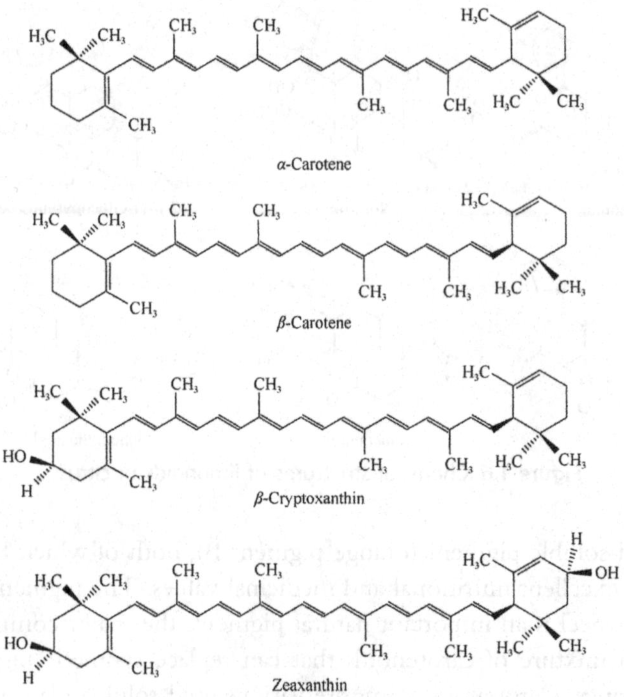

Figure 1.7 Chemical structures of carotenoids in citrus.

annual growth rate is 7–9%. However, only 600 tons of β-carotene are produced every year. At present, the price of natural β-carotene is 8,000,000 yuan/ton, which is twice that of the synthetic product. The global production of natural β-carotene is only 5–6% of the total required, which obviously suggests that there is room for development.

7. CELLULOSE

Dietary fiber can be divided into soluble dietary fiber (SDF) and insoluble dietary fiber. SDF includes pectin, gum, mucus, and part of cellulose. Insoluble dietary fiber includes cellulose, hemicellulose, and lignin. Approximately 70% of the components of insoluble dietary fiber are linear polysaccharides composed of glucose molecules linked by β-1,4 glycoside bonds with a polymerization degree of approximately 3000–10,000. The structure of cellulose is shown in Fig. 1.8. Cellulose and hemicellulose comprise approximately 50–60% of citrus peel, which makes citrus peel a

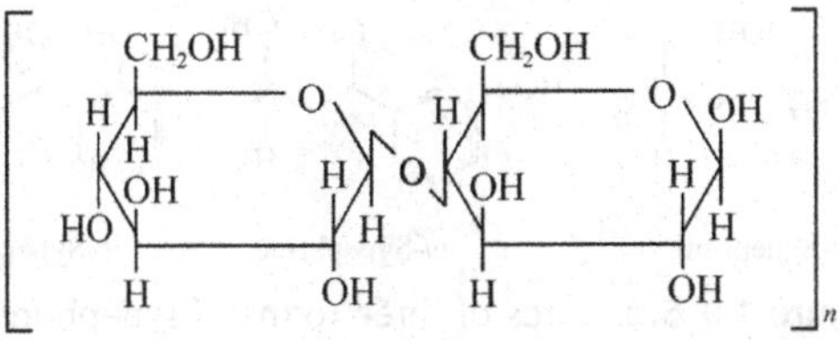

Figure 1.8 The structure of cellulose.

good raw material for the extraction of dietary fiber. Braddock et al. extracted dietary fiber from citrus peel and determined cellulose, hemicellulose, pectin, and lignin contents as 10.8%, 12%, 19.3%, and 6.6% of the total dry weight of citrus peel, respectively. In China, Zhu et al. analyzed the dietary fiber and pectin contents of the peel and pulp from lemon, Yuhuan pomelo, Ponkan, navel orange, Satsuma orange, and grapefruit. The results showed that the SDF content of the flesh is slightly higher than that of the outer peel, except in lemon and navel orange, whereas the contents of other components were higher in the peel than in the flesh. Therefore, citrus peel can be used as a good source of dietary fiber.

The SDF content of citrus peel is several or even dozens of times higher than the dietary fiber content of cereals. The US Food and Drug Administration has ruled that high SDF can be used together with a low-saturated fat or low-cholesterol diet to significantly reduce the incidence of cardiovascular diseases. The cellulose from citrus peel can be widely applied in cakes, biscuits, bread, and other foods. It can also be used as an antitackiness agent in foods. Citrus fiber has major advantages over other dietary fiber products because it contains vitamin C, Ca, K, and other minerals, as well as flavonoids at high concentrations. Therefore, citrus fiber is worthy of development and utilization, with broad market prospects.

8. SYNEPHRINE

Synephrine is naturally present in citrus plants from the Rutaceae family, and it is a sympathomimetic alkaloid with the formula of $C_9H_{13}NO_2$. Synephrine has three different isomeric forms, ie, *p*-synephrine, *m*-synephrine, and *o*-synephrine, as shown in Fig. 1.9; *p*-synephrine occurs naturally in citrus plants from the Rutaceae family, and the highest *p*-synephrine content is found in the fruitlets. Fruit maturation may result in a gradual decline in the *p*-synephrine content. In addition, the *p*-synephrine content may vary among different varieties. Studies have shown that

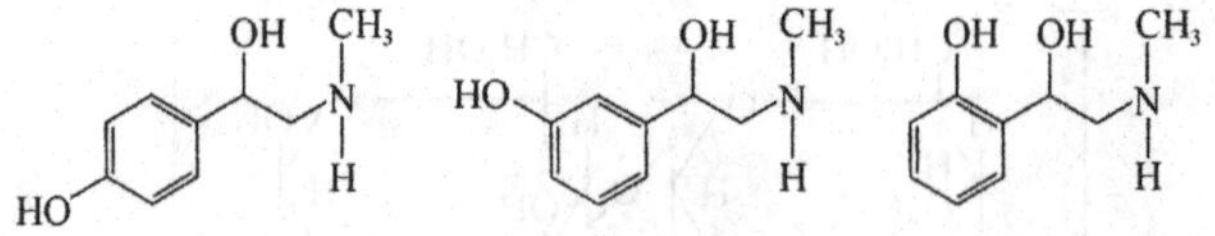

Figure 1.9 Structures of three forms of synephrine.

p-synephrine contents are approximately 0.20–0.27 mg/g in citrus flesh, 53.6–158.1 μg/L in citrus juice, and 1.2–19.8 mg/g in dry fruitlets.

The major component of adrenergic agonists is *p*-synephrine, and it has been documented in pharmacopoeia from Northern Europe and a Germany. It can shrink blood vessels, increase blood pressure, and dilate bronchi. It can also be used as a mild stimulant. Thus, it can be used in clinical treatments of bronchial asthma, hypotension, collapse, shock, and orthostatic hypotension; to improve metabolism; and in treatments of digestive problems such as indigestion, swelling of the liver, phlegm, and ptosis. In foreign countries, *p*-synephrine is utilized for the treatment of obesity by increasing calorie consumption and fat oxidation, as well as mitigating mild depression caused by obesity. At present, *p*-synephrine is widely used in the pharmaceutical, food, beverage, and other industries. The increased market demand for synephrine has doubled its value.

9. OTHER FUNCTIONAL COMPONENTS

In addition to the major types of functional components in citrus peel, there are many other components such as coumarin, acrylic acids, and glycolipids. Coumarin is found at high levels in some fruits and vegetables, and its anticancer activity has been explored. It has been confirmed that coumarin effectively induces the detoxification enzyme system, degrades carcinogens, and inhibits cancer complications at the initial stage.

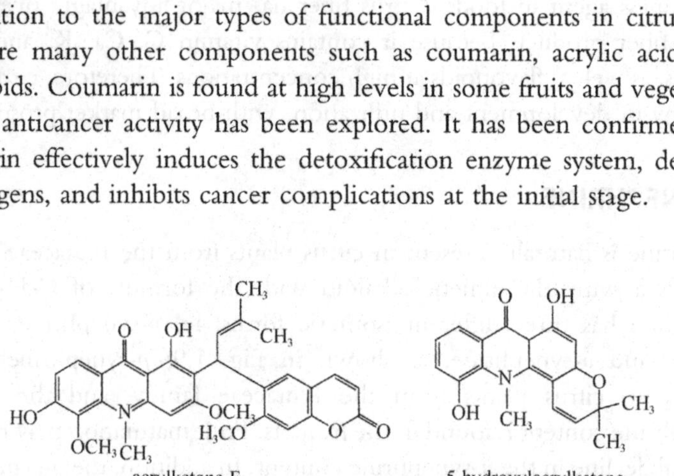

Figure 1.10 Chemical structures of acrylketone components of citrus.

A class of alkaloids called acrylketones has been isolated from the root bark of citrus plants. Two of these acrylketones, acrylketone-F and 5-hydroxylacrylketone, have higher inhibitory activities during the development stage of cancers, and their structures are shown in Fig. 1.10. These compounds also have many other pharmacological functions including antiviral effects. Thus, the functional components of citrus require further exploration, and they are expected to attract greater research attention.

CHAPTER 2

Methods for Determining the Functional Components of Citrus Peel

Contents

1. Chemical Analysis: Determination of Pectin Using a Gravimetric Method 16
   1.1 Principle 16
   1.2 Reagents 16
   1.3 Sample Handling 16
      1.3.1 Fresh Samples 16
      1.3.2 Grinding 17
   1.4 Extraction of Pectin 17
      1.4.1 Extraction of Water-Soluble Pectin 17
      1.4.2 Extraction of Total Pectin 17
   1.5 Calculation 18
   1.6 Notes 18
2. Spectral Analysis: Determination of Total Flavonoids in Citrus Peel by $AlCl_3$-UV-Visible Spectroscopy 18
   2.1 Principle 19
   2.2 Standards and Reagents 19
   2.3 Instruments 19
   2.4 Analytical Procedures 19
      2.4.1 Sample Preparation and Extraction 19
      2.4.2 Establishment of the Standard Curve 19
      2.4.3 Colorimetric Determination 20
   2.5 Determination of Total Flavonoids in Samples 20
   2.6 Precision 20
      2.6.1 Repeatability 20
      2.6.2 Reproducibility 20
3. Determination of Multi-Methoxyflavones in Citrus Peel by High-Performance Liquid Chromatography (NY/T 2336-2013) 20
   3.1 Principle 21
   3.2 Reagents and Materials 21
   3.3 Instruments and Equipment 21
   3.4 Analytical Procedures 22
      3.4.1 Sample Preparation 22
      3.4.2 Extraction and Purification 22
      3.4.3 Instrument Reference Conditions 22
      3.4.4 Standard Curve 23

Comprehensive Utilization of Citrus By-products
ISBN 978-0-12-809785-4
http://dx.doi.org/10.1016/B978-0-12-809785-4.00002-2

3.4.5 Determination 23
3.4.6 Blank Test 23
3.5 Calculation 24
3.6 Precision 24
3.6.1 Repeatability 24
3.6.2 Reproducibility 24
4. Determination of Aromatic Components in Citrus Peel by GC–MS 24
4.1 Extraction of Volatile Oils From Citrus Peel 24
4.2 Headspace Solid Phase Microextraction Conditions 25
4.3 GC–MS Conditions 25
4.4 Data Processing 25
5. LC–MS (HPLC–MS) 29
6. HPLC-Nuclear Magnetic Resonance Spectroscopy 29

There are many functional components in citrus peel, and various methods are available for their determination. The major determination methods include chemical analysis, spectrophotometry, liquid chromatography (LC), gas chromatography–mass spectroscopy (GC–MS), and liquid chromatography-tandem MS (LC–MS)/MS.

1. CHEMICAL ANALYSIS: DETERMINATION OF PECTIN USING A GRAVIMETRIC METHOD

1.1 Principle

The sample is treated with 70% ethanol to precipitate pectin, which is followed by successive washes using ethanol and ether for removing the soluble sugar, fat, and coloring materials. The residues are extracted using acid and water to obtain total and water-soluble pectin. After saponification, sodium pectate is generated, which is then acidified with acetic acid to generate pectic acid. Calcium salt is added to produce the precipitate calcium pectate. Finally, the precipitate is dried and weighed.

1.2 Reagents

Alcohol, phenol, sulfuric acid, hydrochloric acid, acetic acid, and calcium chloride.

1.3 Sample Handling

1.3.1 Fresh Samples

Fresh samples weighing 30–50 g are sliced into small pieces. The samples are placed in a 500 mL Erlenmeyer flask containing 99% ethanol and subjected to reflux in a water bath for 15 min. After refluxing, the samples are cooled and filtered using a Buchner funnel. The residues are transferred to a mortar and ground slowly. During grinding, 70% hot ethanol is added drop by drop. After cooling and filtering, the procedure is repeated until the filtered solution does not exhibit a reaction that indicates sugars (tested using the phenol—sulfuric acid method). Finally, the residues are subjected to washing with 99% ethanol to facilitate dehydration, washing with ether to remove any lipids and pigments, and drying to remove the ether.

1.3.2 Grinding

The dried samples are ground to obtain a fine powder after passing through a 60-mesh sieve. A sample weighing approximately 5—10 g is placed in a beaker and dissolved with 70% hot ethanol, which is coupled with vigorous stirring to extract any sugars. The suspended solution is filtered, and the procedure is repeated until no saccharide reaction is observed. The residues are subjected to sequential washes with 99% ethanol and ether, with a final drying step to remove the latter.

1.4 Extraction of Pectin

1.4.1 Extraction of Water-Soluble Pectin

The residues obtained are suspended in 150 mL of water in a 250-mL beaker, which is heated and kept boiling for 1 h, where any lost water is replenished continuously. After cooling, the solution is transferred to a 250-mL volumetric flask and water is added to a final volume of 250 mL. Following shaking and filtration, the filtered solution is collected as the water-soluble pectin extract.

1.4.2 Extraction of Total Pectin

1. Next, 150 mL of 0.05 mol/L hydrochloric acid is heated to boiling and used to dissolve the residues in the Buchner funnel, which are then transferred to a 250 mL Erlenmeyer flask and refluxed in a boiling water bath for 1 h. After cooling, the suspension is transferred to a 250-mL volumetric flask, and two drops of methyl red indicator are added to the solution.
2. Sodium hydroxide (0.5 mol/L), a neutralizer, is added until the final volume is attained, mixed well, and filtered, and the filtered solution is collected as the total pectin extract.
3. Determination of pectin: In total, 25 mL of the extract (approximately 25 mg of calcium pectate) is transferred to a 500-mL beaker, and 100 mL of 0.1 mol/L sodium hydroxide is added with stirring. After

remaining at room temperature for 0.5 h, 50 mL of 1 mol/L acetic acid is added, followed by stirring for 5 min at room temperature. While stirring, 25 mL of 1 mol/L calcium chloride solution is added slowly. The mixture is left for 1 h, then heated to boiling for 5 min. After filtering through a filter paper that has been dried to constant weight (or a G2 vertical melting crucible), the hot mixture is washed with hot water until no chloride ions are detected (tested with 10% silver nitrate solution). The precipitate and the filter paper are placed in a weighing bottle and dried to constant weight at 105°C in an oven (they can be placed directly in an oven if using a G2 vertical melting crucible).

1.5 Calculation

$$X(\%) = \frac{0.9233 \times (m_1 - m_2) \times V}{V_1 \times m} \times 100\%,$$

where m_1 is the mass of the calcium pectate and the filter paper or vertical melting crucible (g); m_2 is the mass of the filter paper or vertical melting crucible (g); m is the mass of the sample (g); and 0.9233 is the transformation coefficient from calcium pectate to pectic acid.

The empirical formula of calcium pectate is $C_{17}H_{22}O_{11}Ca$, with a calcium content of 7.67% and a pectic acid content of 99.23%.

1.6 Notes

1. The immersion of fresh specimen slices in ethanol can inhibit the pectinase activity.
2. Sugars can be tested using the phenol—sulfuric acid method. Approximately 1 mL of the sample is placed in a test tube, and 1 mL of 5% phenol solution and 5 mL of sulfuric acid are added sequentially to the test tube. After mixing well, the solution turns brown, which indicates the presence of sugar in the test sample.
3. The calcium pectate precipitate mixes readily with other gel substances, which may result in low selectivity with this determination method.

2. SPECTRAL ANALYSIS: DETERMINATION OF TOTAL FLAVONOIDS IN CITRUS PEEL BY $AlCl_3$-UV-VISIBLE SPECTROSCOPY

Spectroscopy is an established method that is based on the selective light absorption of a test material, and this can be used for qualitative and quantitative analyses. Among the methods used for the detection of functional components from citrus, ultraviolet spectrophotometry and fluorescence spectrophotometry are typically selected for the rapid determination of pectin, flavonoids, limonoids, carotenoids, and total polyphenolic compounds.

2.1 Principle

Flavonoids can form a complex with aluminum ions, which results in a color change reaction. A quantitative analysis is possible based on comparisons with standard samples.

2.2 Standards and Reagents

Rutin, ethanol, methanol, and $AlCl_3$ (analytical grade).

2.3 Instruments

a. UV–Vis spectrophotometer
b. Analytical balance with a scale of 0.01 and 0.00001 g
c. Tissue blender
d. Ultrasonic cleaner

2.4 Analytical Procedures

2.4.1 Sample Preparation and Extraction

The pulverized citrus peel powder is dried fully at 55°C and then passed through a 60-mesh sieve. Next, 2.00 g of the dried powder is transferred to a 100-mL volumetric flask and dissolved in 20 mL of 95% ethanol. The solution is subjected to ultrasonic treatment at 42 kHz for 45 min. The ultrasonic-treated solution is collected by filtration, and the residues are dissolved in 20 mL of 70% ethanol for another 45 min with ultrasonic treatment. The ultrasonic-treated, filtered solutions from both cycles are combined and mixed, and then made up to a final volume of 50 mL with 70% ethanol. Next, 20 mL of this mixture is centrifuged at 3500 r/min for 10 min, followed by 5 mL of the supernatant that is measured accurately and diluted to 50 mL.

2.4.2 Establishment of the Standard Curve

The rutin standard (0.0077 g) is weighed and dissolved in 30% ethanol with ultrasonic treatment. The solution is made up to a final volume of 100 mL with 30% ethanol and mixed well for future use.

Aliquots of 0, 1.0, 2.0, 3.0, 4.0, and 5.0 mL of the 0.077 mg/mL rutin standard solution are measured, made up to final volumes of 5.0 mL with 30% ethanol, and then mixed with 4.0 mL of 1% $AlCl_3$ solution. Each standard sample is made up to a final volume of 10 mL using 30% ethanol, followed by shaking and standing for 10 min.

2.4.3 Colorimetric Determination

The standard rutin solutions are subjected to a colorimetric assay at 415 nm, and the absorbance measurements are used to establish a standard curve. The regression equation is as follows: $Y_{415} = 23.032X - 0.0025$ ($R^2 = 0.9993$). In addition, the absorbance of a 1.0 mL volume of each test sample is measured using the same method.

2.5 Determination of Total Flavonoids in Samples

The total flavonoid content of each sample is calculated as follows: $X\ (\%) = C \times V_1 \times V_2 \div V_3 \div m \times 100\%$, where C is the concentration of flavonoids in the sample solution (mg/mL); V_1 is the volume of the sample solution (mL); V_2 is the total volume of the sample solution (mL); V_3 is the volume used for the color reaction (mL); and m is the mass of the sample (mg).

2.6 Precision

2.6.1 Repeatability

To determine the repeatability of the test conditions, two independent measurements must be made where the absolute difference should not exceed 10% of the arithmetic mean.

2.6.2 Reproducibility

To determine the reproducibility of the test conditions, two independent measurements must be made where the absolute difference should not exceed 10% of the arithmetic mean.

3. DETERMINATION OF MULTI-METHOXYFLAVONES IN CITRUS PEEL BY HIGH-PERFORMANCE LIQUID CHROMATOGRAPHY (NY/T 2336-2013)

High-performance liquid chromatography (HPLC) has been widely used for analytical purposes in the past 10 years, ie, for accurate qualitative and quantitative analyses. The sample pretreatment procedure required for HPLC is simple, and the wide range of available separation columns, and mobile phase types and ratios, as well as short analysis times and high sensitivity, explain its rapid development and widespread popularity. Each type of functional component in citrus can be separated and detected by HPLC, mainly using a reverse phase C_{18} column (250 mm × 4.6 mm, particle size of filling materials = 5 μm, with a chemically bonded stationary phase). The most common detector is a UV detector. In recent years, photodiode array detectors (diode array detector [DAD] or photo-diode array [PAD]) have also attracted much attention because they can overcome the shortcomings of ordinary UV detectors and acquire chromatographic and spectral information at the same time, thereby providing a wealth of information that facilitates qualitative and quantitative analyses of samples.

3.1 Principle

Methoxyflavones in the test samples are subjected to ultrasonic vibration extraction in an organic solvent, followed by purification, dilution to a set volume, microporous membrane filtration, and HPLC determination using an external standard method.

3.2 Reagents and Materials

Unless stated otherwise, the reagents used in the analysis are of pure analytical grade and the water is of the GB/T 6682 level.

a. Methanol (CH_3OH, CAS No. 67-56-1), HPLC grade

b. Acetic acid [volume fraction ψ (CH_3COOH) = 0.5%]: 5 mL of acetic acid is diluted to 1000 mL

c. Nobiletin standard (CAS No. 478-01-3; purity >96.0%)

d. Tangeretin standard (CAS No. 481-53-8; purity >96.0%)

e. Sinensetin standard (CAS No. 2306-27-6; purity >96.0%)

f. Preparation of methoxyflavone stock standard solution: In total, 5.00 mg (accurate to 0.01 mg) of sinensetin, tangeretin, and nobiletin standards are each dissolved separately in methanol and diluted to a volume of 50 mL. The final concentration of the prepared standard solutions is 100 mg/L. All of the standard solutions are stored at −50°C for future use.

3.3 Instruments and Equipment

a. HPLC equipped with a UV detector

b. Analytical balance with a scale of 0.01 and 0.00001 g

c. Tissue blender
d. Ultrasonic cleaner
e. Rotary evaporator
f. Membrane: 0.45 μm, for the organic phase

3.4 Analytical Procedures

3.4.1 Sample Preparation

The fresh fruit peel is cut into small pieces by quartering and homogenized using a tissue blender. The same method is used to process the seeds.

3.4.2 Extraction and Purification

For each sample, two separate subsamples of 1.00 g (accurate to 0.01 g) are weighed and dissolved in methanol in a 50-mL centrifuge tube. The final volume is made up to 10.00 mL and then subjected to ultrasonic treatment at 50°C for 30 min. After centrifugation at 10,000 r/min for 10 min, the supernatant is collected and the residues are subjected to the same extraction procedure on two more occasions. The supernatants are combined and made up to a final volume of 50 mL. The prepared solution is filtered through a 0.45-μm filter membrane and stored for future use.

3.4.3 Instrument Reference Conditions

Column: C_{18} column (particle size = 5 μm, 4.6 mm × 250 mm) or a column with equivalent performance
Column temperature: 35°C
Injection volume: 10 μL
Detection wavelength: 330 nm
Mobile phase: the gradient elution program is shown in Table 2.1.

Table 2.1 Gradient elution program for the mobile phase

Time (min)	Flow rate (L/min)	Phase A (acetic acid) %	Phase B (methanol) %
	1.00	60	40
5	1.00	35	65
15	1.00	35	65
16	1.00	10	90
20	1.00	10	90
21	1.00	60	40
25	1.00	60	40

3.4.4 Standard Curve

The stock standard solutions are diluted with methanol to produce final concentrations of 0.02, 0.10, 0.50, 2.5, 10.0, 20.0, and 30.0 mg/L as the standard working solutions, as described in Section 4.3. The standard curves for sinensetin, tangeretin, and nobiletin are plotted with the concentration on the x-axis and the corresponding peak area on the y-axis, and the standard linear regression equations are obtained.

3.4.5 Determination

The two test samples are analyzed separately as described previously. Aliquots of 10 μL of the test sample solutions and the corresponding standard working solutions are injected to facilitate quantitative analysis based on the chromatographic peak areas (see Fig. 2.1). The response values for sinensetin, tangeretin, and nobiletin are in the linear determination range.

3.4.6 Blank Test

Methanol solutions without test samples are determined according to the same procedures.

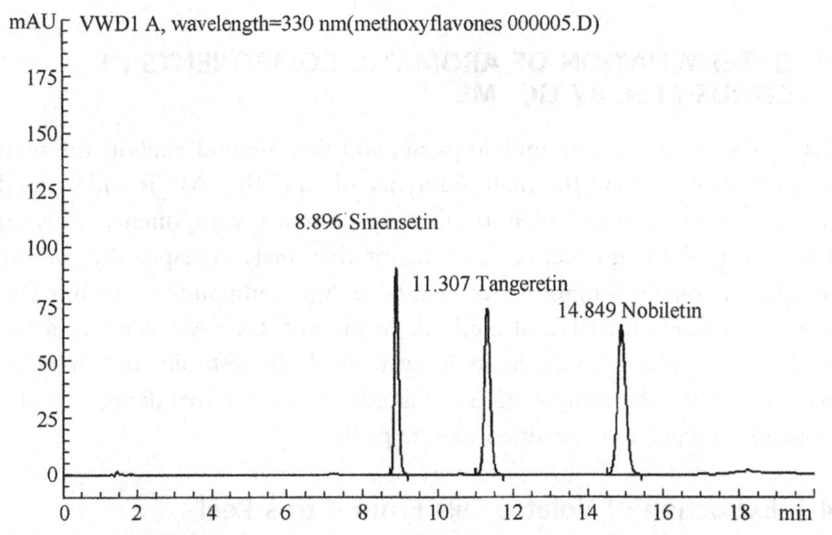

Figure 2.1 Chromatographic plots for 5.00 mg/L sinensetin, tangeretin, and nobiletin.

3.5 Calculation

Sinensetin, tangeretin, and nobiletin contents of the test samples are calculated as the mass fraction ω (mg/kg) using the following formula:

$$\omega = \frac{\rho \times V \times 1000}{m \times 1000} \times n \ldots\ldots\ldots\ldots\ldots\ldots,$$

where ρ is the mass concentration of sinensetin, tangeretin, or nobiletin in a test sample (mg/L); V is the final volume of the test sample solution (mL); m is the test sample mass (g); and n is the dilution factor.

The results are calculated and expressed to three significant figures.

3.6 Precision

3.6.1 Repeatability

To determine the repeatability of the test conditions, two independent measurements must be made where the absolute difference should not exceed 10% of the arithmetic mean.

3.6.2 Reproducibility

To determine the reproducibility of the test conditions, two independent measurements must be made where the absolute difference should not exceed 10% of the arithmetic mean.

4. DETERMINATION OF AROMATIC COMPONENTS IN CITRUS PEEL BY GC—MS

GC—MS uses gas as the mobile phase, and this method exploits the high resolution of GC and the high sensitivity of MS. GC—MS is widely used for the separation and identification of complex components. It is an effective tool for qualitative and quantitative analyses, especially for the analysis of volatile compounds or readily derived compounds, which makes it an ideal modern analytical method. At present, GC—MS is used for the analysis of aromatic components in citrus both domestically and overseas, particularly for the analysis of essential oils in satsuma mandarin, pomelo, ponkan, orange, and sun-dried orange peels.

4.1 Extraction of Volatile Oils From Citrus Peel

After crushing the dried citrus peel and passing through a 60-mesh sieve, 100 g of the citrus peel powder is weighed and transferred to a 2000-mL round-bottom flask. An appropriate amount of water is added to the flask, and the sample is allowed to soak for several hours. The samples are then subjected to reflux, where another 250 mL round-neck flask containing 100-mL anhydrous ether and zeolites is also connected. The citrus peel samples are heated first on the left and the ether is heated on the right when the refluxed liquid begins to flow into it. Both sides are then boiled for 4 h to extract the volatile oils. The mixture of ether and volatile oils is then combined with an appropriate amount of anhydrous sodium sulfate, and the sample is placed in a refrigerator overnight. On the following day, the mixture is condensed using an "Oldshow" column, and a pale yellow transparent oily liquid with a citrus scent is obtained; the yield of the citrus oil is 1.2%.

4.2 Headspace Solid Phase Microextraction Conditions

Solid phase microextraction (SPME)—aged conditions: when it is used for the first time, the extraction head must be primed in the GC injection port (under nitrogen) at 270°C for more than 2 h. Aging is also required during later use for 30 min at 270°C to ensure the removal of any adsorbed volatile components. An aliquot of 7 mL of the essential oil extracted from citrus peel is loaded into a 15 mL vial and subjected to headspace adsorption for 40 min at 60°C and desorption at 250°C for 2 min.

4.3 GC—MS Conditions

GC conditions: capillary column, DB 25 column (60 m × 0.132 mm i.d.; film thickness, 1 μm); column temperature program: initial temperature of 40°C maintained for 1 min, which is then increased at a rate of 5°C/min until the temperature reaches 130°C. The temperature is then increased at a rate of 8°C/min until it reaches 200°C. The final temperature is increased at a rate of 12°C/min until it reaches 250°C, and this temperature is maintained for 7 min. Vaporization chamber temperature: 250°C; carrier gas: He; flow rate: 0.8 mL/min; injection mode: splitless injection mode in the initial stage for 2 min, followed by an injection split ratio of 12:1.

MS conditions: electrospray ionization-MS (ESI-MS), electron energy of 70 eV, filament emission current of 200 μA, ion source temperature of 200°C, interface temperature of 250°C, and a mass scanning range of 33—450 amu.

4.4 Data Processing

The experimental data are processed using Xcalibur software. Unknown compounds are identified using both the NIST library (107,000 compounds) and Wiley library (320,000 compounds, Version 6.0). The identification

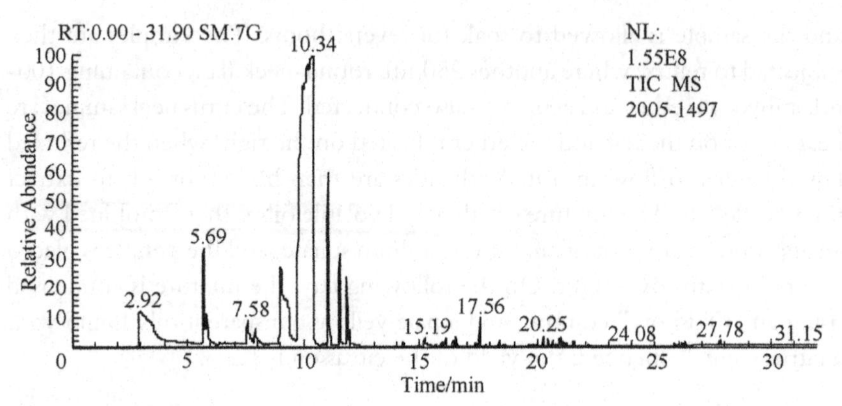

Figure 2.2 GC–MS total ion chromatograph of the essential oil from satsuma mandarin.

results can only be utilized if the matching degree between positive and negative results is more than 800 (the maximum value is 1000).

One of the shortcomings of headspace SPME is the small amount of impure peaks in the trace, which are produced by the siloxane head. Therefore, these impure peaks should be removed during the MS search.

A total ion chromatogram of volatile chemical components of the essential oil from Satsuma mandarin is shown in Fig. 2.2, which was obtained using the experimental procedures described above.

The components of the essential oil from satsuma mandarin shown in Fig. 2.2 were separated and identified using the conditions described previously. The chemical compositions were determined by searching the standard NIST and Wiley libraries and by MS combined with the retrieval of relevant literature using the GI 701 BA ChemStation. The relative contents of the volatile compounds were determined using the peak area normalization method, as shown in Table 2.2.

As shown in Table 2.1, the volatile compounds extracted from Satsuma mandarin mainly comprise hydrocarbon compounds, which account for 96.56%, followed by alcohols, aldehydes, ketones, esters, and phenolic compounds, which account for 0.78%, 0.59%, 0.54%, 0.17%, and 0.13%, respectively. The hydrocarbons comprise unsaturated olefinic compounds (84.85%), including isolimonene (25.92%), 2-(1-methyl acetyl)-bicyclo [2,2] heptane (19.82%), 4-methyl-1-(1-methylethyl)-cyclohexene (15.81%), β-myrcene (8.59%), (E)-3,7-dimethyl-1,3,6-octatriene (6.41%), α-pinene (2.74%), and methyl (1-methylethyl) benzene (2.11%). The alkyne compounds mainly comprise 2-ethyl-octen-3-acetylene (9.20%). The alcohols

Table 2.2 Chemical composition of essential oil extracted from satsuma mandarin

No.	Retention time (min)	Compound name	Formula	Area (%)
Hydrocarbons 96.56				
1	5.69	Alpha-pinene	$C_{10}H_{16}$	2.74
2	5.81	2-Methyl-5-(1-methylethyl)-bicyclo[3,10]hex-2-ene	$C_{10}H_{16}$	1.31
3	7.58	6,6-Dimethyl-2-methylene-bicyclo[3,11]heptane	$C_{10}H_{16}$	1.30
4	7.92	4-Methylene-1-(1-methylethyl)-bicyclo[3,10]hexane	$C_{10}H_{16}$	0.70
5	9.00	Beta-myrcene	$C_{10}H_{16}$	8.59
6	10.00	Isolimonene	$C_{10}H_{16}$	25.92
7	10.17	2-(1-Methylethylidene)-bicyclo[2,2]heptane	$C_{10}H_{16}$	19.82
8	10.34	4-Methyl-1-(1-methylethenyl)-cyclohexene	$C_{10}H_{16}$	15.81
9	10.40	2-Ethyl-1-octen-3-yne	$C_{10}H_{16}$	9.20
10	11.08	(E)-3,7-dimethyl-1,3,6-octatriene	$C_{10}H_{16}$	6.41
11	11.54	Methyl(1-methylethyl)-benzene	$C_{10}H_{14}$	2.11
12	11.83	1-Methyl-4-(1-methylethylidene)-cyclohexene	$C_{10}H_{16}$	1.46
13	15.19	1-Methyl-4-(1-methylethenyl)-benzene	$C_{10}H_{12}$	0.40
14	15.94	4-Ethyl-3-(1-methylethenyl)-1-(1-methylethyl)-cyclohexene	$C_{15}H_{24}$	0.08
15	16.41	Alpha-copaene	$C_{15}H_{24}$	0.10
16	17.37	Beta-cubebene	$C_{15}H_{24}$	0.06
17	18.39	Beta-elemene	$C_{15}H_{24}$	0.11
18	18.53	Caryophyllene	$C_{15}H_{24}$	0.03
19	19.64	*Cis*-2,6-dimethyl-2,6-octadiene	$C_{10}H_{18}$	0.03
20	19.79	(E,E,E)-2,6,6,9-Tetramethyl-1,4,8-cycloundecatriene	$C_{15}H_{24}$	0.04
21	20.41	Germacrene-D	$C_{15}H_{24}$	0.13
22	20.94	E,E-alpha-farnesene	$C_{15}H_{24}$	0.13
23	21.12	Delta-cadinene	$C_{15}H_{24}$	0.05
24	22.14	*Cis*-calamenene	$C_{15}H_{22}$	0.03

Continued

Table 2.2 Chemical composition of essential oil extracted from satsuma mandarin—cont'd

No.	Retention time (min)	Compound name	Formula	Area (%)
Alcohols 0.78				
1	17.56	L-linalool	$C_{10}H_{18}O$	0.49
2	17.81	1-octanol	$C_8H_{18}O$	0.04
3	18.65	4-terpineol	$C_{10}H_{18}O$	0.03
4	20.25	L-alpha-terpineol	$C_{10}H_{18}O$	0.16
5	21.23	(R)-3,7-dimethyl-6-octen-1-ol	$C_{12}H_{20}O$	0.06
Aldehydes 0.59				
1	14.27	Nonanal	$C_9H_{18}O$	0.09
2	16.10	3,7-dimethyl-6-octenal	$C_{10}H_{18}O$	0.16
3	16.52	*n*-decanal	$C_{10}H_{20}O$	0.18
4	19.96	(Z)-3,7-dimethyl-2,6-octadienal	$C_{10}H_{16}O$	0.03
5	20.73	E-citral	$C_{10}H_{16}O$	0.04
6	21.51	4-(1-methylethenyl)-1-cyclohexene-1-carboxaldehyde	$C_{10}H_{14}O$	0.09
Esters 0.17				
1	20.64	Neryl acetate	$C_{12}H_{20}O_2$	0.09
2	21.07	2,6-octadien-1-ol,3,7-dimethyl-acetate	$C_{12}H_{20}O_2$	0.07
3	27.78	Diethyl phthalate	$C_{12}H_{14}O_4$	0.10
Phenols 0.13				
1	25.80	Thymol	$C_{10}H_{14}O$	0.03
2	26.14	4-ethyl-2,6-dimethyl-phenol	$C_{10}H_{14}O$	0.04
3	26.27	Thymol	$C_{10}H_{14}O$	0.06
Ketones 0.54				
1	11.94	1,3-cycloheptadien-1-ylmethylketone	$C_9H_{12}O$	0.54

mainly include L-linalool (0.49%) and L-α-terpineol (0.16%). The ketone compounds mainly include 1,3-cycloheptadiene-1-methyl ketone (6.33%). The aldehyde compounds mainly comprise decanal (0.18%) and 3,7-dimethyl-6-octanal (0.16%).

5. LC–MS (HPLC–MS)

Liquid chromatography–mass spectrometry (LC–MS) is a high-tech detection method based on high-capacity chromatographic separation and high-sensitivity MS, which is mainly used for the analysis and detection of trace components. LC–MS can be used for the accurate qualitative and quantitative analyses of bioactive components. The HPLC–MS technique is extensively used for the detection of functional components from citrus. At present, the most commonly used interface and ionization technologies are ESI and atmospheric pressure chemical ionization.

For example, Anagnostopoulou et al. used HPLC-DAD-ESI-MS to detect flavonoids in citrus peel. The DAD detector can analyze a single sample at different wavelengths simultaneously, which may reflect the absorption characteristics of specific materials. Seventeen different types of flavonoids (including two types of unknown compounds) have been detected in citrus peel using this method. The compounds detected mainly comprise methoxyflavones, *C*-glycoside flavonoids, *O*-glycoside flavonoids, *O*-glycoside flavanones, flavanols, phenolic acids, and their derivatives.

6. HPLC-NUCLEAR MAGNETIC RESONANCE SPECTROSCOPY

HPLC-nuclear magnetic resonance (NMR) is a new analytical tool used to determine functional components in modern research, which exploits the advantages of both LC–MS and LC–MS–MS. This method has an irreplaceable role in the structural identification of components in a mixture. In the HPLC-NMR technique, HPLC facilitates effective separation and NMR provides a wealth of information about the structure of compounds, such as the stereochemistry. HPLC-NMR can detect structural information related to a compound, but its sensitivity is determined by the concentration of the compound. Therefore, this technique is currently limited to compounds with relatively low molecular weights. However, if conventional methods have low efficiency and are ineffective for detection or identification, the advantages of HPLC-NMR can be exploited, ie, fast and efficient detection. HPLC-NMR can be used for the separation and structural identification of crude extracts that contain trace components. Therefore, it may have important roles in the analysis of functional components in various contexts.

CHAPTER 3

Extraction Processes of Functional Components From Citrus Peel

Contents

1. Classification and Extraction Processes of Essential Oils From Citrus Peel 33
 1.1 Major Citrus Oils 33
  1.1.1 Orange Oil 33
  1.1.2 Mandarin Oil 34
  1.1.3 Bitter Orange Oil 34
  1.1.4 Tangerine Oil 34
  1.1.5 Lemon Oil 34
  1.1.6 Lime Oil 34
  1.1.7 Bergamot Oil 35
  1.1.8 Grapefruit Oil 35
 1.2 Preparation of Citrus Oil 35
  1.2.1 Cold Milling Method 35
  1.2.2 Cold-Pressed Method 35
  1.2.3 Water Distillation Method 36
 1.3 Key Features of the Cold Milling Process for Producing Essential Oils 36
  1.3.1 Cleaning 36
  1.3.2 Dermabrasive Oil Mill 36
  1.3.3 Filtration 37
  1.3.4 Centrifugation 37
  1.3.5 Standing 38
  1.3.6 Condensation and Terpene Removal 38
  1.3.7 Packaging and Preservation 38
 1.4 Key Features of the Extraction Process of Cold-Pressed Oil 38
  1.4.1 Material Selection 38
  1.4.2 Soaking in Lime 38
  1.4.3 Cleaning 39
  1.4.4 Crushing 39
  1.4.5 Quality Standards 39
 1.5 Extraction Process of d-Limonene 39
2. Extraction Process of Pectin From Citrus Peel 41
 2.1 Key Operating Points of High Methoxyl Pectin Extraction 42
  2.1.1 Raw Material Selection 42
  2.1.2 Pretreatment of Raw Materials 42
  2.1.3 Acidic Extraction 42
  2.1.4 Filtration 43

Comprehensive Utilization of Citrus By-products
ISBN 978-0-12-809785-4
http://dx.doi.org/10.1016/B978-0-12-809785-4.00003-4

2.1.5 Decoloration 43
2.1.6 Condensation 43
2.1.7 Precipitation 43
2.1.8 Drying 44
2.1.9 Standardization 44
2.2 Preparation of Low Methoxyl Pectin 44
2.2.1 Alkali Preparation Method 44
2.2.2 Acidification Method for Pectin Preparation 44
2.2.3 Enzymatic Method for Pectin Preparation 45
2.2.4 Resin and Phosphate Method 46
2.2.5 Membrane Separation 48
2.3 Quality Standards of Pectin 48
3. Extraction Processes of Flavonoids From Citrus Peel 49
3.1 Hesperidin Extraction Process and Operating Points 49
3.1.1 Raw Materials 49
3.1.2 Lime Treatment 49
3.1.3 Squeezing, Precipitation, and Filtration 50
3.1.4 Neutralization and Keeping Warm 50
3.1.5 Cooling and Precipitation 50
3.1.6 Separation 50
3.1.7 Refining 50
3.1.8 Quality Standards 51
3.2 Naringin Extraction Process and Operating Points 51
3.2.1 Extraction 51
3.2.2 Purification Process 51
4. Extraction of Limonoid and its Analogs 51
4.1 Limonin Extraction 52
4.1.1 Comprehensive Extraction Process 52
4.1.2 Dichloromethane Extraction Process 52
4.1.3 Ultrasonic Extraction Process 52
4.1.4 Supercritical CO_2 Extraction Process 52
4.2 Separation and Purification 53
5. Extraction Process of Synephrine From Citrus Peel 53
5.1 Technological Process 53
5.2 Key Operating Points 54
5.2.1 Sample Treatment 54
5.2.2 Soaking in Water Overnight 54
5.2.3 Percolation Conditions 54
5.2.4 Separation and Purification 54
5.3 Product Quality Indexes 54
6. Extraction Process of Edible Cellulose Powder 54
6.1 Technological Process 55
6.2 Key Operating Points 55
6.2.1 Grinding and Soaking in Acid Solution 55
6.2.2 Immersion in Water 55
6.2.3 Bleaching, Filtration, and Drying 55
6.2.4 Crushing and Sieving 55

6.3 Product Quality Indexes 55
7. Extraction Process of Seed Oil 56
7.1 Technological process 56
7.2 Key Operating Points 56
7.2.1 Cleaning and Drying of Raw Materials 56
7.2.2 Grinding 56
7.2.3 Oil Extraction 56
7.2.4 Solvent Extraction 57
7.2.5 Refining 57
7.2.6 Quality Standards 58

This chapter discusses the extraction processes of the major functional components from the by-products of citrus peel, such as citrus oil, pectin, flavonoids, limonoids, and synephrine.

1. CLASSIFICATION AND EXTRACTION PROCESSES OF ESSENTIAL OILS FROM CITRUS PEEL

Citrus, including orange, red orange, bitter orange, lemon, lime, bergamot, and grapefruit, are important fruit and spice plants. These fruits have a thick cortex that contains large amounts of aromatic oils. Cold-pressed oil can be obtained by milling or pressing, and it is subjected to steam distillation to acquire distilled oil. In general, the cold-pressed oil has the highest quality.

The essential oil obtained from citrus is dominated by terpenes, sesquiterpenes, higher alcohols, aldehydes, and esters. The first two types of hydrocarbons account for 95% of the total essential oil, whereas the oxygen-containing compounds account for less than 5%. However, the latter are the major sources of essential oil aroma.

1.1 Major Citrus Oils

1.1.1 Orange Oil

The cold-pressed or cold-milled oil is usually an orange liquid, whereas the distilled oil is light yellow. The cold-milled and cold-pressed oils are usually derived from fresh orange peel, which means that it has the aroma of an orange, although the aroma of the distilled oil is weak. The aroma quality is correlated with changes in the aroma components. The oxygen-containing aldehydes, alcohols, and esters have stronger aromas, whereas the more abundant terpenes have a lighter aroma. Orange-flavoring oil is extensively used in orange soda and orange juice, as well as in toothpaste, cigarettes, perfumes, and detergents.

1.1.2 Mandarin Oil

Mandarin oil can be divided into cold-pressed and distilled oils. The cold-pressed mandarin oil is an orange or orange-red liquid, the distilled mandarin oil is light yellow with a fragrant wax lipid or slight amine aroma, and it is not as sweet as orange oil.

1.1.3 Bitter Orange Oil

Cold-pressed bitter orange oil is a yellow or light yellow liquid. It has a particularly fruity odor, which is similar to lemon or grapefruit. Despite its slightly bitter taste, it can still be used to make orange wine.

1.1.4 Tangerine Oil

Cold-pressed tangerine oil is a red, orange-red, or dark brown liquid, whereas distilled tangerine oil is lighter in color. Its aroma is similar to that of orange oil with a fresh orange aroma, which means that it is often used in perfume and cologne.

1.1.5 Lemon Oil

Lemon oil is also divided into cold-pressed and distilled oils. The cold-pressed lemon oil is a yellow or yellow-green liquid, whereas the distilled lemon oil is a pale yellow liquid. Lemon oil has a fresh and elegant lemon aroma, as well as clear pinene and terpinene aromas. Its distinctive aroma is derived from aliphatic aldehydes, terpene alcohols, terpene alcohol-acetic esters, and terpenes (limonene, pinene, and terpinene). Lemon oil is the major lemon flavor agent used in lemon-based drinks, toothpastes, shampoos, detergents, and creams. It is also used widely in other types of non-lemon flavor essence, especially in chypre, fougère, cologne, and lavender perfumes, where it provides the well-known "top note."

1.1.6 Lime Oil

Lime oil can be obtained from the whole fruit or peel by the cold-pressed or distillation method. Different sources and processing methods lead to major variations in the aroma of lime oil. The aroma differences depend on the oil components. Higher levels of terpineol, borneol, and terpinene confer a heavier cool fragrance, whereas aliphatic aldehydes, neral, geraniol, geraniol, and linalool have the opposite effect in producing fresh and sweet fruity scents. The aroma of cold-pressed lime oil is better than that of distilled lime oil, which is also the case for lemon oil. In general, the aroma of fresh lime oil is fresh and cool, and it can be used in drinks, candy, and food flavors.

1.1.7 Bergamot Oil

Bergamot oil can be obtained from lemon peel by the cold-pressed or distillation method. The cold-pressed bergamot oil is a green or yellow-green liquid, whereas the distilled bergamot oil is a pale yellow liquid. The aroma of top bergamot oil is similar to that of lemon, which is followed by an orange blossom scent with a unique aroma because it contains high levels of linalool and linalyl acetate. Bergamot oil is an important natural spice, which is widely used in various fragrances, especially perfumes, cologne, and toilet water.

1.1.8 Grapefruit Oil

Grapefruit oil can be obtained from the whole fruit and peel by the crushing or distillation method. The cold-pressed grapefruit oil is a yellow or yellow-green liquid, whereas the distilled grapefruit oil is pale yellow. The aroma of grapefruit oil is similar to that of orange with a fresh and sweet flavor. Grapefruit oil can be used as the major flavoring agent to develop beverages with grapefruit flavors. In addition, it is often combined with other citrus oils to produce everyday fragrances, especially in perfumes and colognes.

1.2 Preparation of Citrus Oil

The most common preparation methods for citrus oil are cold milling, cold pressing, and water distillation.

1.2.1 Cold Milling Method

Cold-milled oil is obtained by breaking oil cells by mechanical means without chemical or heating treatments. The cold-milled oil can be flushed out by spraying with water, which is followed by centrifugation. This method produces good quality and high-value essential oils.

Whole citrus fruit ⟶ Washing ⟶ Mechanically milled oil from citrus peel ⟶ Oil–water ⟶ Filtration ⟶ High-speed centrifugation ⟶ Still stratification ⟶ Reduced-pressure filtration ⟶ Citrus essential oil ⟶ Removal of terpenes ⟶ Edible citrus fragrance.

1.2.2 Cold-Pressed Method

To extract oils by cold pressing, the fruit peel is usually soaked in lime solution to stiffen the fruit peel and pressed to disrupt oil cells and release the oil. The cold-pressed oil can then be obtained by spraying with water and then centrifugation.

Raw material selection ⟶ Soaking ⟶ Washing ⟶ Spraying press ⟶ Filtration ⟶ Centrifugation ⟶ Cold-pressed oil.

1.2.3 Water Distillation Method

Distilled oils can be obtained from crushed peel, the residues of cold-milled or cold-pressed oil, and wastewater. Many of the important fragrant components of essential oils, such as alcohols, aldehydes, and esters, are lost by heating, which means that distilled essential oils have the lowest quality.

Citrus peel ⟶ Crushing ⟶ Distillation ⟶ Oil–water separation ⟶ Distilled citrus oil.

1.3 Key Features of the Cold Milling Process for Producing Essential Oils

1.3.1 Cleaning

The fresh fruits are washed in flowing water to remove any dirt and leaves. The cleaned fruits are then soaked in 0.5% sodium carbonate solution for 1–3 min and dried by natural drainage.

1.3.2 Dermabrasive Oil Mill

Traditional oil mills include the Avicenna and Fuana types. The Avicenna type is also known as a drum-grinding mill. Because of its simple manufacturing method and low cost, this type of mill is used widely in China. In recent years, the Fuana type was introduced into China because of the increasing production of concentrated juice. This mill comprises parallel steel drums, which are covered with spikes on their surfaces. During juice preparation using an FMC whole fruit miller, the essential oil extracted from the fruit peel has the same quality as the cold-milled oil. The oil production process using an Avicenna mill is performed as follows.

First, raw materials are divided into large, medium, and small fruits because the rotation speed and milling duration depend on fruit size.

Large fruits: 150–160 r/min for 75–80 s

Medium fruits: 140–150 r/min for 70–75 s

Small fruits: 130–140 r/min for 65–70 s

The fruits are placed in the oil drum of the mill in batches at specific time intervals. After completing the milling process, the door is opened to discharge the raw materials, and the milled fruits are released. During milling for oil production, water is sprayed to draw the oil into a hopper. Spraying water ensures the timely collection of the essential oil produced because it inhibits the reabsorption of oil by spongy tissues in the pericarp.

Maintaining the turgor of fruit peel parenchyma cells can result in the outflow of essential oil from oil cells during cell disruption, which can prevent the loss of essential oil. The amount of water sprayed should be correlated with the fruit-feeding quantity and the separation capacity, thereby controlling the oil production process. In general, approximately 50 kg of water is sprayed per ton of citrus fruit. The sprayed water can also be recycled, although it may become cloudy and sticky because it contains increasing amounts of pectin or sediments. Thus, some of the water should be released and replaced with fresh water, with pH adjusted to 7.5–8.0. The released spraying water usually contains essential oil, which can be recovered by distillation.

1.3.3 Filtration

In general, a shaking sieve can be used to filter the oil–water mixture. To reduce the burden during milling, the oil–water mixture can be passed through an 80–100-mesh sieve to remove the peel and viscous mash. A large amount of essential oil can be recovered from the residues by distillation.

1.3.4 Centrifugation

The separation process is usually divided into two steps. The first separation step uses an automatic slagging machine to isolate an oil–water emulsion with a high essential oil content. The second separation step uses an essential oil separator. Because of the different densities of the essential oils from diverse citrus fruit varieties, the watershed ring of the oil–water separator needs to be selected according to the specific raw material, such as red-orange oil, lemon oil, orange oil, and grapefruit oil, which require watershed ring diameters of 110–113, 116, 110–116, and 113 mm, respectively. When the centrifuge reaches the normal operating speed, the filtrate can be fed into the machine with a controlled flow rate while spraying appropriate amounts of water. The flow rate must not be too high or too low. The flow rate of the disc separator without the slagging function is usually controlled at 12–24 L/s, while that of the automatic slagging disc separator is generally 18–20 L/s. After the oil–water mixture has been separated, a small amount of oil remains in the drum of the separator; therefore, the centrifuge needs to be operated for another 2–3 s without loading. The remaining oil and oil clay can also be diluted with water for additional separation, or distillation can be used for the further recovery of the essential oil.

1.3.5 Standing

The essential oil separated from the citrus peel usually contains small amounts of water, wax, and other impurities. Thus, it needs to be stored in a refrigerator at 8°C for 6 days to precipitate the impurities. The clear essential oil in the upper level can then be collected using a siphon. The citrus peel oil obtained is a yellow oily liquid with a sweet orange aroma. This type of oil is lighter than water and insoluble in water. However, it can be dissolved in ethanol at 7–10 times the volume.

1.3.6 Condensation and Terpene Removal

To reduce storage and transport costs, the essential oil needs to be condensed. Condensation can also remove a certain amount of the water-insoluble terpene compounds. Limonene (the major component of terpene olefins) is unstable in light and heat; therefore, it is easily oxidized by atmospheric oxygen to produce the odor compound carvone, which results in lower-quality essential oil. At present, distillation and solvent methods are the major condensation processes. The distillation method can facilitate separation based on the different boiling points of the essential oil components. The solvent method allows the separation of terpene olefins and other compounds based on differences in their solubility in various solvents.

1.3.7 Packaging and Preservation

Light, temperature, moisture, air, and impurities, such as iron or other metal ions, can lead to the oxidative deterioration of essential oil. Therefore, essential oils should be packaged in aluminum barrels, white iron barrels, or iron barrels with paint on the inner surfaces. The containers should be filled completely and sealed well under low-temperature storage conditions.

1.4 Key Features of the Extraction Process of Cold-Pressed Oil

1.4.1 Material Selection

The raw material used to produce cold-pressed oil is fruit peel, from which it is necessary to remove mildew and blanched peel.

1.4.2 Soaking in Lime

Soaking with lime can change the pectin in citrus peel into insoluble calcium pectate, which makes the citrus peel harder, thereby facilitating crushing, filtration, and oil–water separation. In general, 1.5% lime can be used for soaking the citrus peel. To attain a pH of 12, the citrus peel is soaked for 6–10 h. The preferred soaking duration is 6–8 h for citrus peel and 8–10 h for orange peel. The ratio of peel relative to soaking solution is 1:4. The soaking process is optimal when the citrus peel retains a fresh color, is slightly hard with no white core, is flexible, and its oil cells have a high-powered jetting force. If a white core is still present, the soaking duration needs to be extended. In contrast, a yellow or dark yellow color suggests excessive soaking. Inadequate or excessive soaking can affect the yield and quality of the essential oil. Lime can be reused 2–3 times, but pH must be maintained at 12. The lime used can be retained for the extraction of hesperidin.

1.4.3 Cleaning

Cleaning the fruit peel with fresh water can remove lime and other impurities, as well as reduce pH to less than 9. In the absence of cleaning, the quality of the essential oil will be affected. After cleaning, the fruit peel should be dried under natural conditions.

1.4.4 Crushing

A horizontal spiral presser can be used for crushing. A spiral roller can forcibly transfer the peel and provide pressure at the same time, leading to the destruction of oil cells. In addition, spraying water can transfer the essential oil into the hopper. The amount of water sprayed should match the capacity of the peel feeder and separator. The water sprayed should also be recycled and its pH needs to be maintained, as described for the cold milling process.

The methods used for filtration, isolation, purification, refining, and product packaging are the same as those used for the cold-milled oil.

1.4.5 Quality Standards

The quality standards for cold-milled and cold-pressed oils are shown in Table 3.1.

1.5 Extraction Process of d-Limonene

d-limonene is a citrus by-product, which was originally referred to as citrus oil. d-limonene can be distilled from pressed juice, waste vapor, and peel oil suspension, or recovered from condensed citrus oil. It is the major edible component of the essential oil obtained from citrus peel. In addition, it can be produced from the by-products of condensed citrus oil.

Table 3.1 Quality standards of essential oils obtained from citrus peel

Index	Essential oil type			
	Cold-milled orange oil	**Cold-milled lemon oil**	**Cold-pressed orange oil**	**Cold-pressed sweet orange oil**
Color	Yellow or lemon color	Slight yellow and yellow-green color	Orange-red or brown	Yellow
Aroma	Sweet orange aroma	Lemon aroma	Citrus aroma	Sweet orange aroma
Refractive index (20°C)	1.4742	1.474–1.478	1.4739	1.4754
Specific rotation (20°C)	+97.00	+65.00	+92.23	+91.00
Density (20°C)	0.8472	0.856–0.861	0.8504	0.8542
Acid value (mg/g)	0.47	—	0.66	1.54
Ester content (mg/g)	1.06	—	1.12	2.45
Aldehyde content[a] (%)	1.90	2.5	1.46	9.23
Nonvolatile residues (%)	3.73	—	3.72	9.23

[a]Aldehyde amount: hydratropaldehyde is the major aldehyde present in orange essential oil. Citralin is the major aldehyde present in lemon and orange essential oils. The volume factions represent the aldehyde contents of the corresponding essential oil.

d-limonene is a by-product of the different cold-pressed oils or essential oils obtained during condensation. Distillation can be used to condense essential oil and for producing limonene. High temperatures can result in the degradation of essential oil components; therefore, distillation is usually performed under reduced pressure. d-limonene with a high purity (more than 99%) is obtained by distillation in a highly efficient manner. The distillation process starts under low vacuum conditions to collect components with low boiling points. d-limonene can be collected as vacuum increases. Distillation must be performed with care, and d-limonene and other impurities such as myrcene and octanol with similar boiling points

Table 3.2 Characteristics of crude d-limonene

Index	Characteristic value	Index	Characteristic value
Relative density (20°C)	0.84	Myrcene content (%)	1.8
Refractive index (20°C)	1.471	α-Pinene content (%)	0.6
Specific rotation (20°C)	+96.7	Sabinene content (%)	0.4
Flash point (°C)	46	Aldehyde content (%)	0.4
Boiling point (°C)	165	Low boiling point octanal (%)	0.1
d-Limonene (GC) (%)	94.6	Color	Colorless to pale yellow
High boiling point components content (%)	2.3		

Data resources for crude d-limonene, from Ye X. Citrus processing and comprehensive utilization. Beijing: China Light Industry Press; 2005.

can be collected. If a specific level of purity is required, diluted NaOH or a carbonyl addition agent (eg, hydroxylamine hydrochloride) can be used to remove aldehydes during distillation, which ensures that product purity is 99.5%. Activated silica can be used as an adsorbent to remove impurities with carbonyl groups or its oxidized products. Oxidizing agents, such as hypochloric acid can be used to remove reducing compounds in d-limonene. After purification, d-limonene is readily oxidized in the air. If it is not protected, d-limonene can be converted to crude d-limonene with a purity of approximately 95–96%. The characteristics of crude d-limonene are shown in Table 3.2.

2. EXTRACTION PROCESS OF PECTIN FROM CITRUS PEEL

Pure pectin is generally white or pale yellow with no fixed solubility and melting point (m.p.), soluble in water, and slightly soluble in cold water. It can be almost completely dissolved in 20 fold volume of water to form a viscous liquid. It is acidic to litmus and insoluble in ethanol and glycerol or other organic solvents. It is stable under the mild acidic condition but easy to decompose under the alkaline condition.

In the network structure model of the cell wall, crystal cellulose and hemicellulose are connected by hydrogen bonds to form a network of fiber and hemicellulose. Hydrophilic pectin and a small amount of structural protein are filled in the space of the mesh structure. The mechanism underlying the pectin extraction process is complex, and is accompanied by the component degradation of plant cell walls. Cell wall degradation is a comprehensive chemical reaction involving the breakage of secondary bonds and covalent bonds in each component, breakage of glycosidic bonds in components, and degradation of polysaccharides. Moreover, polysaccharide degradation forms the major part of this reaction.

2.1 Key Operating Points of High Methoxyl Pectin Extraction

The production process of high methoxyl pectin is as follows:

Dried citrus peel → Smashing → Washing → Extraction → Filtration → Reduced vacuum condensation → Precipitation → Drying smashing pectin.

2.1.1 Raw Material Selection

The white cortex and central column of citrus contain 20–40% pectin (dry base) and are good raw materials for pectin extraction. On the basis of species analysis, lemon and lime peels are the best, with an average pectin content of 35.5%. Pectin content is 25% in orange peel, 20% in grapefruit peel, and 10% in loose-skin orange peel. Under the optimal extraction conditions, the gel degree reaches up to 300–350 for lemon peel, 250–300 for grapefruit peel, and 100–180 for loose-skin orange peel.

Besides species, the pectin content is correlated with maturation of raw materials. The higher maturation of raw materials will result in a lower content and quality of pectin.

2.1.2 Pretreatment of Raw Materials

After removing of spoiled moldy citrus peel and washing, the chosen citrus peel is cut into small pieces (dimension, 1–2 cm^2). Dual-step countercurrent washing with warm water at 40°C will reduce the content of soluble solids to approximately less than 1.5%. It is better to control water content at 8–10% during the drying process.

2.1.3 Acidic Extraction

During industrial production of pectin, acidic extraction remains the major extraction method. The pectin-containing dried citrus peel is mixed with water and hydrochloric acid. The ratio of raw material to acid solution is 1:8. The pH is controlled at 2–3 for dissolving pectin, and then the extraction of pectin is conducted for several hours at 50–100°C.

2.1.4 Filtration

The extracted solution is heated to approximately 90°C, and the spiral presser supplied with a filter (pore size, 0.5 mm) is used to extract pectin. The crude pectin is filtered and centrifuged at high speed to remove impurities.

2.1.5 Decoloration

Activated carbon of up to 0.3–0.5% is added to the extract. With stirring at 55–60°C for 20–30 min, the extract is discolored, and then 1–1.5% diatomaceous earth is added to the extract for facilitating filtration. After filtration, the transparent and clarified pectin extract is obtained.

2.1.6 Condensation

The transparent and clarified pectin extract is transferred to a vacuum condenser to concentrate at 85 MPa until the content of soluble solids reaches 7–9%. To produce liquid pectin, the pectin extract concentrate is heated at 70°C and pH is adjusted to 3.5–4 using ammonia or sodium carbonate. After condensation, the liquid pectin is sealed in a container and sterilized at 70°C in a water bath for 30 min. Then, the liquid pectin is stored in the refrigerator. In addition, the liquid pectin can be sealed in a barrel with the addition of 0.1% sodium sulfite.

2.1.7 Precipitation

Most pectin is sold in the form of pectin powder. Thus, pectin products should be precipitated from the solution. Ethanol precipitation is the first method used in industrial production of pectin. After 1.5% hydrochloric acid is added to the concentrated industrial pectin extract, 90% ethanol at the same volume (identical to the volume of industrial pectin extract) is added slowly with stirring at an interval of 1–2 min. Upon stirring four or five times, the mixture is allowed to stand for 20 min and pectin is precipitated. To ensure full precipitation of pectin, the ethanol content should be higher than 45–50%. Pectin is obtained through filtration in a spiral presser and washed with 95% ethanol two or three times (0.5 h each time).

2.1.8 *Drying*

The pressed pectin can be dried in a vacuum drying room or a drum dryer until the water content is less than or equal to 8%. The temperature is controlled at 65–75°C in the vacuum drying room but a little higher in the drum dryer. Dried pectin is crushed using a steel mill and passed through a 60 to 80-mesh sieve.

2.1.9 *Standardization*

The gel degree of the pectin powder obtained in the industrial production line is different. Thus, it is subjected to standardized treatment, whereas pectin with a higher gel degree is supplemented with a certain amount of sugar and that with a lower gel degree is supplied with a certain amount of pectin with a higher gel degree. The gel degree of commercially available high methoxyl pectin is usually 150.

2.2 Preparation of Low Methoxyl Pectin

The preparation methods of low methoxyl pectin mainly include alkali, acid, resin and phosphate, and enzymatic preparation methods.

2.2.1 *Alkali Preparation Method*

The pectin concentrate with a pectin content of 4% is placed in a stainless pot. Ammonium hydroxide (0.5 mol/L) is added to the pectin concentrate to adjust pH to 10.5 and kept at room temperature for 3 h. After allowing to stand for 3 h, 95% ethanol at the identical volume (identical to the volume of industrial pectin extract) is added, and an appropriate amount of hydrochloric acid is added to adjust pH to 5. After stirring and allowing to stand for 1 h, the pectin precipitate is obtained after filtration. The pectin precipitate is washed with 50% and 95% ethanol (50% ethanol is to prewash and get pectin easily precipitated; 95% ethanol is to remove the organic compounds more completely) and dried on a baking tray at 50°C in a vacuum dryer to attain a moisture content of 8%. The dried pectin is ground and packaged as the final product. The yield of pectin is approximately 90%.

2.2.2 *Acidification Method for Pectin Preparation*

The pectin concentrate with a pectin content of 3% is placed in a stainless jacketed kettle, and pH is adjusted to 3 with hydrochloric acid at 50°C. Hydrolysis is conducted at this temperature to remove the esters until a viscous mixture is formed. Further hydrolysis is required to reduce the methoxyl content to the desired level. After hydrolysis, ethanol is added to the hydrolysates to precipitate pectin. After filtration under reduced pressure, the pectin precipitate is washed with fresh water to remove acid and then neutralized with diluted alkaline solution. Followed by repeated ethanol precipitation of pectin, the pectin is dried at 65°C in a vacuum dryer until the moisture content is less than or equal to 8%, and then ground to form powder. The dried pectin powder is packaged after passing through a 100-mesh sieve.

2.2.3 Enzymatic Method for Pectin Preparation

The low methoxyl pectin is extracted after degreasing by pectin lipase. Changhe et al. from the Fruit Research Institute in Guangdong Province (1996) has successfully developed an industrial production technology of pectin through enzymatic extraction from pomelo peel. Compared with the traditional alkali and acidification methods for pectin preparation, the enzymatic extraction method has the advantages of convenient operation, high product quality, low energy consumption, and low production cost.

1. Process technology:
   Pomelo → Smashing → Washing → Degreasing → Extraction → Filtration → Precipitation → Salt removing → Ethanol washing → Filtration → Drying → Smashing → Final product.
2. Key operating points
   a. Mincing of raw materials: Raw materials are minced into small pieces of 3–5 mm.
   b. Water washing: The materials are immersed in water at 50°C for 30 min, centrifuged, and rinsed with water two or three times until the eluate is colorless.
   c. Degreasing: An appropriate amount of sodium carbonate is added to activate pectin esterase and degreasing. Degreasing process conditions include a temperature of 50°C, duration of 1 h, pH of 7.0, and sodium carbonate concentration of 7 g/kg. However, as for fresh fruit peel, the sodium carbonate concentration at 25 g/kg dry peel is the best.
   d. Pectin extraction: Hydrochloric acid is added to adjust pH to 1.7–2.0 and pectin extraction at 95°C.
   e. Precipitation: An appropriate amount of $CaCl_2$ is added to precipitate pectin.
   f. Desalination and alcohol washing: Hydrochloric acid and oxalic acid are mixed at a ratio of 1:3 for removing salts, and alcohol washing is performed for several times.

g. Drying and grinding: Vacuum drying at 60°C is performed, and the dried pectin is pulverized as the powder. The gel degree of pectin is 100 ± 5, and the degree of esterification is less than 50%, which meets the requirements of the US food additive standards (Food Chemical Codex).

2.2.4 Resin and Phosphate Method

1. Technological process

a. Extraction of wet materials

Cation exchange resin ↓

Fresh citrus peel → Immersion, enzyme inactivation → Mincing → Rinsing → Hydrochloric acid, phosphate extraction → Filtration, bleaching → Condensation → Ethanol precipitation → Filtration, washing → Drying → Pectin.

b. Extraction of dry materials

Cation exchange resin ↓

Fresh citrus peel → Drying → Smashing → Immersion, enzyme inactivation → Rinsing → Hydrochloric acid, phosphate extraction → Filtration, bleaching → Condensation → Ethanol precipitation → Filtration, washing → Drying → Pectin.

2. Key operating points

a. Resource and pretreatment of raw materials

The scraps from juice or canned food manufacturers are dried and saved for pectin preparation.

b. Drying and grinding

The raw materials are dried at 65°C, pulverized to particles with a required size (particle size 3–5 mm), and then sealed and stored for future use.

c. Pretreatment of cation exchange resin

Dry cation exchange resin is soaked in distilled water overnight or stirred for 2 h to make it fully swollen. The swollen resin is soaked with four volumes of 2 mol/L hydrochloric acid for 1 h or stirred for 0.5 h. The supernatant is discarded and the resin is washed to achieve a neutral status. Then, the resin is treated with 2 mol/L sodium hydroxide as mentioned in the previous method. Finally, the resin is treated with

2 mol/L hydrochloric acid and transformed into hydrogenated resin. After washing to achieve a neutral status, the resin is dried for future use.

d. Soaking and enzyme inactivation

A certain amount of water is added to the raw materials, and an appropriate amount of cation exchange resin is added. The mixture is heated to 85–90°C, and the temperature is held to inactivate the enzyme.

e. Rinsing and drying

The citrus peel subjected to enzyme inactivation is washed several times with hot water at 50–60°C to remove free monosaccharides, aromatic substances, salts, bitter taste, soluble noncolloidal substances, pesticide residues, chemical fertilizers, and other toxic substances, as well as small molecule compounds, such as pigments. After washing, the pomace is dried using a crushing machine.

f. Pectin extraction

The dried pomace is placed in the reactor containing water at 95°C. Hydrochloric acid is added to adjust pH to 2. According to the weight of the extract, 2% sodium hexametaphosphate is added and reacted for 1.5 h.

g. Separation and bleaching

Approximately 2–3% activated carbon is added to the filtrate and incubated at 60°C for 20–30 min to deodorize and remove the odor. The aperture of the filter cloth is selected on the basis of the difficulty degree of filtration and the ash content of final products. Suction filtration is conducted when the solution is still hot. If the viscosity of the filtrate is too high or filtration is too difficult, 3–4% diatomaceous earth can be added as a filter-aiding agent, or filtration is conducted under reduced pressure after stirring. If the citrus peel is cleaned, the extract is clear and transparent and bleaching is not required.

h. Condensation, precipitation, and purification of the extract

The precooled 95% ethanol in the multistrands form is injected into the cooled and concentrated pectin until the ethanol concentration in the filtrate is 50–60%. The mixture is allowed to stand for 30–45 min, and pectin is precipitated. The mixture of ethanol and pectin is filtered, and the residues are washed with acidified 90–95% ethanol. Finally, filtration and rinsing with acidified anhydrous ethanol and dehydration are performed.

i. Drying

The resultant residue cake after filtration is broken up and paved in a thin layer. The pectin is obtained through vacuum or oven drying to attain a moisture content of 7%. After weighing, the yield of pectin is calculated.

j. Calculation of pectin yield

Extraction rate of pectin (%) = weight of the final pectin product/raw product weight × 100%.

2.2.5 Membrane Separation

Membrane separation is the method using ultrafiltration and electrodialysis to purify and condense pectin. Using membrane separation technology, small molecule impurities are separated from the pectin to accomplish pectin condensation at room temperature, which can avoid the decomposition or destruction of active components.

1. Technological process

   Grapefruits after oil milling → Washing → Smashing → Pretreatment, extraction, centrifugation, and filtration → Filtrate → Membrane separation → Supernatant → Evaporating condensation → Spray drying → Packaging → Final products.

2. Description of the main steps

   The citrus fruit peel is used as the raw material (dry raw material after rehydration by soaking for 2–3 h). The raw materials are cooked in boiling water at 100°C for 5–7 min to inactivate enzymes. The enzyme-inactivated raw materials are subjected to repeat rinsing to remove the impurities, such as pigments, bitter-tasting substances, and sugar in raw materials. The rinsed raw materials are minced and soaked in hydrochloric acid solution at pH 2.0. Extraction is conducted under the following conditions: material–liquid ratio of 20:1, extraction temperature of 85–95°C, and extraction duration of 1–2 h. The crude extract of pectin obtained by ultrafiltration is preliminarily concentrated to remove impurities without contribution to gelation and most of acids and inorganic ions. The high-quality pectin is acquired through spray-drying from the extract.

2.3 Quality Standards of Pectin

Technical requirements in GB 25533-2010 *Food Additives, Pectin* are shown in Table 3.3 and Table 3.4.

Table 3.3 Sensory requirements

Items	Requirements	Testing methods
Color	White, pale yellow, light yellow, or light brown	An appropriate amount of the samples is placed in a clean and dry white porcelain dish, and the color and appearance of the samples are examined under natural light
Morphology	Powder	

Table 3.4 Physiochemical indexes

Items	Index	Testing methods
Weight loss after drying, w/%	≤12	GB 5009.3 direct drying[a]
Sulfur dioxide/(mg/kg)	≤50	GB/T 5009.34
Insoluble ash in acid, w/%	≤1	GB 25533-2010A.3 in Appendix A
Total galacturonic acid, w/%	≥65	GB 25533-2010A.4 in Appendix A
Amidation degree (only amidated pectin), w/%	≤25	GB 25533-2010A.4 in Appendix A
Lead (Pb)/(mg/kg)	≤5	GB 5009.12
(Methanol + ethanol + isopropanol)[b], w/%	≤1.0	GB 25533-2010 in Appendix B

[a]Drying temperature and duration are 105°C and 2 h, respectively.
[b]Only nonethanol processed products.

3. EXTRACTION PROCESSES OF FLAVONOIDS FROM CITRUS PEEL

Citrus is rich in flavonoids that are easy to separate and have unique aroma and remarkable pharmacological effects. In the mature fruits, higher contents of flavonoids are found in peel, pit, and pulp, but lower contents are found in the juice, only 1–5%. In this section, the extraction processes of hesperidin and naringin are described.

3.1 Hesperidin Extraction Process and Operating Points

The principle of hesperidin extraction is to dissolve hesperidin in alkaline hot water and to precipitate it in acidic cold water. The recycling process is as follows:

Citrus peel ⟶ Lime treatment ⟶ Smashing, precipitation, and filtration ⟶ Neutralization and keeping warm ⟶ Cooling down and precipitation ⟶ Dehydration ⟶ Mincing ⟶ Crude products ⟶ Refining ⟶ Refined products.

3.1.1 Raw Materials

The raw materials include an immersion solution of citrus peel obtained after preparing canned citrus products or citrus pomace obtained after juicing, as well as wastewater obtained after citrus processing.

3.1.2 Lime Treatment

The citrus peel is cut into small pieces with a size of 0.5 cm. After cleaning, the pieces are soaked in a clear saturated lime solution for 6–12 h. The pH should be more than 11. Each batch of lime solution can treat three batches of raw materials to ensure more hesperidin content in the solution.

3.1.3 Squeezing, Precipitation, and Filtration

The abovementioned soaking solution is pressed using a dual conical presser to obtain a liquid. The liquid together with the lime solution is filtered again. For example, the soaking liquid obtained from citrus oil processing is allowed to stand for 4–5 h to accomplish full precipitation. The supernatant is sucked and filtered through a press or stocking filter to produce a clear filtrate.

3.1.4 Neutralization and Keeping Warm

The supernatant is adjusted to pH 4.0–4.5 using hydrochloric acid and incubated at 60–70°C for 40–50 min.

3.1.5 Cooling and Precipitation

After cooling and allowing to stand for 24–48 h, hesperidin is precipitated at the bottom of the jar.

3.1.6 Separation

The precipitate is obtained from the mother liquor using a siphon, centrifuged, and dried with hot air at 70–80°C until a moisture content of less than or equal to 3% is attained. The crude hesperidin content is approximately 40%. Because the total loss of hesperidin in the raffinate is approximately 40–50%, the residues can be recrystallized by condensation.

3.1.7 Refining

The crude pectin is dissolved in 0.2 mol/L isopropanol containing 50% NaOH to obtain a 2% solution, filtered, adjusted to pH 8.5, and allowed to stand for 48 h. The purified crystals are filtered and dried. For the purity of the final product to be more than 95%, this operation process is repeated several times. In addition, hesperidin can be dissolved in 10% formamide, diluted with water, and recrystallized.

During the extraction of hesperidin, the use of plain iron equipment should be avoided. During long-term storage, the crude product should not be stored in a dry environment to avoid the growth of microorganisms. Because of the low solubility of hesperidin in water, it easily crystallizes on the surface of instruments. Meanwhile, hesperidin hydrolysis generates hesperetin, rhamnose, and glucose, thereby reducing its economic value. Therefore, frequent cleaning of the equipment is highly required.

3.1.8 Quality Standards

The final product is a light yellow powder with a purity of no less than 90% for the first grade, no less than 80% for the second grade, and no less than 70% for the third grade. In addition, the moisture, ash, insoluble solid contents in saturated lime, and soluble material content in chloroform should be no more than 5%, 3%, 1.2%, and 1.5%, respectively.

3.2 Naringin Extraction Process and Operating Points

3.2.1 Extraction

Similar to the extraction of hesperidin, naringin can be extracted in alkaline solution, such as lime or sodium hydroxide. It has more solubility in water than hesperidin. The pH for extraction is usually 9–10. Precipitation and recycling of naringin are similar to those of hesperidin. However, a higher naringin concentration is required for crystallization to ensure better crystals than those of hesperidin. A previous study has also demonstrated that the residues obtained after the extraction of naringin by hot water, rather than that after extraction by alkaline solution, are more suitable as raw materials for the extraction of pectin.

3.2.2 Purification Process

The crude products are dissolved in 8–10% boiling isopropanol for recrystallization. The crystals are needle shaped and can be easily broken.

4. EXTRACTION OF LIMONOID AND ITS ANALOGS

Limonoid and its analogs are abundant in citrus fruits, especially in peel and seeds. The limonoid content of grapefruit seeds is up to 1.5% of the fresh weight of the seeds. Because of their biological activities, such as anticancer, health-promoting, and highly efficient insecticide functions, limonoid and its analogs have gained increasing attention.

Limonoid analogs are of many types. Different growth seasons and different tissues of grapefruits can result in the different types and contents of limonoids. Therefore, the selection of extraction methods or conditions (acidic or alkaline) for limonoids should be suitable for different purposes and different raw materials. For example, ripened fruits have a higher content of limonin glycosides, and fruits during the early development period have a higher content of aglycone. In addition, limonin glycosides usually exist in a pH 7.5 environment and can be easily hydrolyzed in too acidic or too alkaline environments.

4.1 Limonin Extraction

The selection of extraction solvents in classic extraction methods is based on the polarity of components. The common extraction methods of limonin from citrus peel include hot extraction (boiling and refluxing), immersion extraction (percolation and cold immersion), and supercritical extraction methods. The limonoid content of citrus peel is 100–300 mg/kg.

4.1.1 Comprehensive Extraction Process

Currently there are many extraction processes, where the solvents involved are ethyl acetate, acetone, water, acidic water, and acetone–methanol mixture.

The extraction process is as follows:

Citrus peel → Ethyl acetate extraction → Solvent layer → Separation and purification → Residues → *n*-Butanol extraction → Solvent layer → Separation and purification → Limonoids.

4.1.2 Dichloromethane Extraction Process

The optimal extraction process of limonoids from citrus peel includes dichloromethane as the extraction solvent, repeated extraction for two times, a refluxing duration of 2 h, an extraction temperature of 50°C, and a solid–liquid ratio of 1:12.

4.1.3 Ultrasonic Extraction Process

Fresh pomace is used as the raw material and adjusted to pH 4 using acidic water as the extraction solvent. Under the extraction conditions with a material–liquid ratio of 1:20 and extraction temperature of 50°C, the highest extraction rate is achieved.

4.1.4 Supercritical CO_2 Extraction Process

As a modern extraction technology, the supercritical CO_2 extraction process has the following unique advantages. It resolves environmental pollution caused by the solvent, is safe without any reactions with CO_2, prevents CO_2 oxidation, and maintains the stability of the extracted substance. Supercritical fluid CO_2 extraction requires an optimal extraction pressure, extraction temperature, extraction duration, and appropriate entrainer. Supercritical fluid CO_2 extraction can excellently prevent the loss of limonin.

4.2 Separation and Purification

The crude extract has complex components and more impurities. Therefore it needs further separation and purification. It is thus particularly important to determine a reasonable and efficient separation and purification route. Commonly used purification methods include crystallization, extraction, column chromatography, and preparative high-performance liquid chromatography.

5. EXTRACTION PROCESS OF SYNEPHRINE FROM CITRUS PEEL

p-Synephrine is present in most citrus fruits, with a higher content observed especially in young fruits. However, *p*-synephrine content reveals a decreasing trend as the fruits ripen. Meanwhile, the difference in the *p*-synephrine content is also due to the different varieties. Research data show that the synephrine content is in the range of 53.6–158.1 μg/L in orange juice and 1.2–19.8 mg/g in dried orange peel.

At present the extraction methods of synephrine reported in the literature include boiling, methanol–ethanol diluted solution or weak acid solution (diluted hydrochloric acid or diluted phosphoric acid) extraction, percolation extraction, refluxing extraction, ultrasonic-assisted extraction, and microwave-assisted extraction methods. The extraction efficiency of synephrine mainly depends on the extraction methods and solvents used. The separation and purification technologies of synephrine mainly include water extraction and ethanol precipitation, cation exchange resin, the ion-exchange column chromatography–cation exchange resin method, macroporous resin, macroporous ion-exchange resin, and the filtration membrane process. In this section, the percolation method using hydrochloric acid and the cation exchange resin extraction process have been described.

5.1 Technological Process

Citrus peel, *Citrus aurantium* powder ⟶ Soaking in water overnight ⟶ Filtration ⟶ Filtered cake ⟶ Percolation using hydrochloric acid ⟶ Filtrate condensation ⟶ Synephrine extract ⟶ Cation exchange resin ⟶ Synephrine.

5.2 Key Operating Points

5.2.1 Sample Treatment

Citrus peel and *C. aurantium* are dried and crushed for passing through a 40-mesh sieve.

5.2.2 Soaking in Water Overnight

Approximately fivefold volume of water is added to the coarse powder of the sample for wetting, which will reduce the diffusion resistance during dissolution. After being soaked overnight, the sample is filtered.

5.2.3 Percolation Conditions

The percolation tube is filled with hydrochloric acid solution to avoid the direct application of hydrochloric acid on the cake for reducing the dissolution of more impurities. The hydrochloric acid concentration, liquid–material ratio, and flow rate of hydrochloric acid during the percolation process are 0.02 mol/L, 1:9, and 5 mL/min, respectively.

5.2.4 Separation and Purification

Through static adsorption and elution experiments, strong acid cation exchange resin is used for the purification of synephrine. This type of resin has higher adsorption capacity to synephrine and reveals better desorption performance. The column separation and purification conditions include a synephrine concentration of 1 mg/mL for loading, solvent pH of 2.4–2.8, sample flow rate in the column of 6 mL/min, 5% aqueous ammonia used as the eluent, and eluent flow rate of 3 mL/min.

5.3 Product Quality Indexes

The final product is a brownish-yellow to yellow-brown powder with a characteristic smell of *C. aurantium* and bitter taste. Granularity [rate of passing through a 80-mesh sieve (%)], ≥90; weight loss rate during drying (%), ≤6.0; residue after ignition (%), ≤8.0; and synephrine content (%), ≥10.0.

6. EXTRACTION PROCESS OF EDIBLE CELLULOSE POWDER

The amount of citrus peel annually discarded by citrus processing factories is very high. Although many processing industries have used citrus peel as the raw material for producing citrus peel oil, citrus pigments, pectin, and hesperidin, residues obtained after extracting (approximately greater than or equal to 60% of citrus peel in weight) have not been well utilized. According to the chemical analysis, these residues are mainly composed of cellulose and hemicellulose.

6.1 Technological Process

Citrus peel → Smashing → Soaking in acid solution → Soaking in water → Bleaching → Rinsing → Filtration → Vacuum drying → Smashing → Sieving products.

6.2 Key Operating Points

6.2.1 Grinding and Soaking in Acid Solution

The natural dried citrus peel is crushed to a granule size of 1−2 mm and soaked in 0.02 mol/L hydrochloric acid. After stirring at room temperature for 0.5 h, filtration is conducted and the residues are saved for future use.

6.2.2 Immersion in Water

The pretreated citrus peel is placed in water at a ratio of 1:20 according to weight. After adjusting pH to 2.0 and incubating at 85−90°C, extraction is conducted for 1 h with stirring. After filtration under reduced pressure, the filtrate is the pectin extract.

6.2.3 Bleaching, Filtration, and Drying

After filtration, the residues are soaked in water at 50−60°C and rinsed until a neutral status is achieved. The residues are treated with 5% hydrogen peroxide at 20−30°C for 10−15 min at pH 5−7 for discoloration. Then, the treated samples are sequentially washed with water and 40−50% ethanol, filtered under reduced pressure, and subjected to vacuum drying.

6.2.4 Crushing and Sieving

The dried cellulose is pulverized into powder by passing through a 200-mesh sieve. The final product is a pale yellow, edible cellulose powder.

6.3 Product Quality Indexes

The product quality of edible crystalline cellulose meets national standards with a crude fiber content of 48.6%, carbohydrate content (cellulose and dry base) of 98.2%, heavy metals (according to Pb) of 8.0 mg/kg, a weight loss of less than 8.0% during drying, pH of 6.5, residue of 0.08% after ignition, and water-soluble substance of 0.13%.

7. EXTRACTION PROCESS OF SEED OIL

The oil content of fresh citrus seeds reaches up to 20–30%. Crude seed oil can be used as the raw material for the manufacturing of soap and sulfonated oils, as well as in the textile industry. After further refining, it is a yellow, bitter, and odor-free liquid, which is similar to aromatic olive oil and can be used as cooking oil. After hydrogenation, it can also be used for preparing margarine and can be used as a flavoring agent in beverages after bromide treatment.

7.1 Technological process

Citrus seeds ⟶ Washing ⟶ Drying ⟶ Pressing during heating or solvent extraction ⟶ Refining ⟶ Final products.

7.2 Key Operating Points

7.2.1 Cleaning and Drying of Raw Materials

After processing, the seeds are cleaned and dried to a moisture content of 8–10% and then stored for oil extraction. The removal of impurities can improve the yield and quality of the oil. Milling will be facilitated because the water content is adjusted and temperature is increased by drying.

7.2.2 Grinding

The seeds are crushed using a drum machine to form flakes. The seed coats are weeded using a winnowing system. Grinding can result in the destruction of tissue cells, increase in the surface area, convenient heat absorption, and water adjustment. Heating can facilitate the solidification and further pressing of proteins. However, formation of too thin flakes or powder should be avoided during pressing because then air penetration will be difficult. In that case, a small amount of water should be added according to the rescue strategy.

7.2.3 Oil Extraction

Orange oil can be extracted by hot-pressed, cold-pressed, and solvent extraction methods. In general, the cold-pressed method produces good quality oil, whereas the hot-pressed method produces oil at a high yield. Therefore, both methods can be applied combinatorially. The residues can also be extracted by the solvent method.

In the cold-pressed method, the oil can be obtained through direct squeezing of seeds. However, in the hot-pressed method, the seeds must first be completely steamed or fried to adjust the water content, improve product temperature, decrease oil viscosity, accelerate protein coagulation, and facilitate oil release. The current hot-pressed method requires the following conditions: water vapor heating for 1 h, pressure control at 0.05 MPa, and temperature control at 103–110°C. The pressure should be uniform, and the rolling ring should be loose without sticky phenomenon. The pressed cake should be brown and have a slightly oily surface.

During pressing, raw materials must be fed evenly. If using an intermittent pressing machine, the required pressure should be increased gradually.

7.2.4 Solvent Extraction

After squeezing, the residues are extracted with the solvent-squeezed part of the residual oil. The commonly used solvents are low boiling point compounds benzene, ethanol, or ether. After extraction, the oil is collected and separated by distillation. The solvents can be recycled for future use.

7.2.5 Refining

Orange oil contains free fatty acids, phosphates, pigments, limonoids, and hesperidin. Crude oil is red with a bitter taste and special odor. Therefore, it must be refined through saponification and bleaching before consumption. The technological process includes refining of the crude oil by alkali treatment, saponification and precipitation, filtration under reduced pressure, drying of essential oil, vacuum deodorization, and bleaching.

1. Refining of crude oil by alkali treatment

   The crude oil is heated with steam, and NaOH is added to this oil according to the content of free acids. The alkali-treated crude oil is milky after stirring. The treated crude oil is incubated at 50°C for 10–15 min with slow stirring for saponification. Saponification can remove free fatty acids and bitterness.

2. Saponification and precipitation

   The oil naturally precipitates and the saponified product can be isolated or removed through centrifugation.

3. Filtration under reduced pressure

   Approximately 4–6% activated carbon is added as the filtration-aiding agent. Followed by full stirring, the impurities and bitterness are removed by filtration under reduced pressure.

4. Drying of essential oil

   The oil is heated to 105–110°C for 20–30 min to remove the water in oil.

5. Deodorization

During the drying process, the agitated air can remove the odor and moisture. Similarly, the vacuum method can also be used for deodorization. In this deodorization process, the hot oil is transferred to the vacuum tank and heated to 150–156°C for vacuum deodorization.

6. Bleaching

Citrus seed oil contains flavonoids and other impurities; therefore, it has a darker color. To improve the color, 2–3% bone char can be added and filtered when it is hot. After filtering, the oil is pale yellow and transparent.

7.2.6 Quality Standards

Citrus seed oil is liquid at room temperature. It also contains 30–60 mg vitamin E in each 100 g sample. Especially, α-tocopherol is the majority component.

Physicochemical properties of citrus seed oil and rapeseed oil are compared in Table 3.5.

Table 3.5 Physicochemical properties of citrus seed oil and rapeseed oil

Index	Citrus seed oil	Citrus seed refined oil	Rapeseed oil
Refractive index (25°C)	1.4676	1.4675	1.4710–1.4735
Density (g/cm^3, 20°C)	0.9191	0.9185	0.910–0.920
Acid value (mg KOH/g oil)	2.29	0.71	Grade I < 0.20, Grade II < 0.30, Grade III < 1.0, Grade IV = 3.0
Iodine value (gI/100 g)	—	121.59	94–120
Saponification value (mg KOH/g oil)	196.33	195.12	168–181
Unsaponification value/(%)	1.33	1.26	≤2

CHAPTER 4

Isolation and Structural Identification of Flavonoids From Citrus

Contents

1. Technological Process for the Separation and Purification of Flavonoids From Citrus Peel 60
   1.1 Qualitative Test Methods of Flavonoids 60
   1.2 Extraction of Flavonoids From Citrus Peel 60
   1.3 Ethanol Extraction Treatment 61
   1.4 Separation of Flavonoids 62
2. Structural Identification of Flavonoids From Citrus Peel 62
   2.1 HPLC—MS 62
   2.2 NMR Technique 63
   2.3 Identification of the Flavonoid Monomer 63

Extraction, isolation, and structural identification of monomer compounds from natural plants are important aspects of natural product research. They are important for (1) discovering the bioactive components of medicinal plants and exploring the mechanisms underlying their clinical efficacy; (2) facilitating clarification of the structure—activity relationship of natural drugs; (3) and providing the evidence and direction to plant classification and physiological studies.

Separation and purification methods of flavonoids include column chromatography, high-performance liquid chromatography (HPLC), and centrifugal thin-layer chromatography (TLC). Column chromatography is one of the most commonly used methods for the separation and purification of flavonoids, which include polyamide column chromatography, silica gel column chromatography, and Sephadex column chromatography. In addition, HPLC separation of flavonoids is reported extensively.

At present, structural elucidation of flavonoids is mainly focused on spectroscopy including mass spectroscopy (MS), ultraviolet spectroscopy, and nuclear magnetic resonance (NMR) spectroscopy. During studies on the aglycone structure of flavonoids, MS displays an obvious advantage

Comprehensive Utilization of Citrus By-products
ISBN 978-0-12-809785-4
http://dx.doi.org/10.1016/B978-0-12-809785-4.00004-6

because flavonoid aglycone has a strong molecular ion peak. However, it becomes slightly difficult for studies of flavonoid glycosides because of the occurrence of fragments on the connected glycons. Flavonoids have absorption bands at both 240–280 nm and 300–400 nm. The former comes from the electronic transition of atoms in the B ring, while the latter comes from the electronic transition of atoms in the A ring. Upon the addition of different chemical agents, such as sodium methoxide, sodium ethoxide, aluminum chloride, boric acid, and sodium acetate, both absorption bands of flavonoid aglycone exhibit a shift due to the change in the structure of flavonoid aglycone. Therefore, mobile agents can be used to determine the substitution locations of aglycones. NMR techniques for elucidating the flavonoid structure include ^{1}H NMR and ^{13}C NMR as well as DEPT-NMR. In Hunan Institute of Agricultural Products, the crude extract of flavonoids obtained from Satsuma mandarin by ultrasonic-assisted extraction is used as the starting material to separate and purify the flavonoid monomer using silica gel chromatography, polyamide column chromatography, Sephedex LH-20 column separation, and purification methods. The purified monomer is subjected to structural identification through spectral scanning, HPLC–MS, and NMR.

1. TECHNOLOGICAL PROCESS FOR THE SEPARATION AND PURIFICATION OF FLAVONOIDS FROM CITRUS PEEL

The technological process for the extraction, separation, and purification of flavonoids from citrus peel is shown in Fig. 4.1.

1.1 Qualitative Test Methods of Flavonoids

TLC analysis of the sample is performed using prefabricated high-performance silica TLC plates through the development of the chloroform–acetic acid–ethyl acetate solvent mixture according to the specified ratio.

Developed TLC plates are sprayed with ammonia, and flavonoids and flavonols reveal yellow bands under visible light. In contrast, under ultraviolet light, flavonoids exhibit bright yellow bands and flavonols reveal bright yellow, yellow-green, or green bands.

1.2 Extraction of Flavonoids From Citrus Peel

In total, 12 kg citrus peel is crushed to form powder and placed in a 60-L plastic bucket. The bucket is filled to the brim with 95% (v/v) ethanol

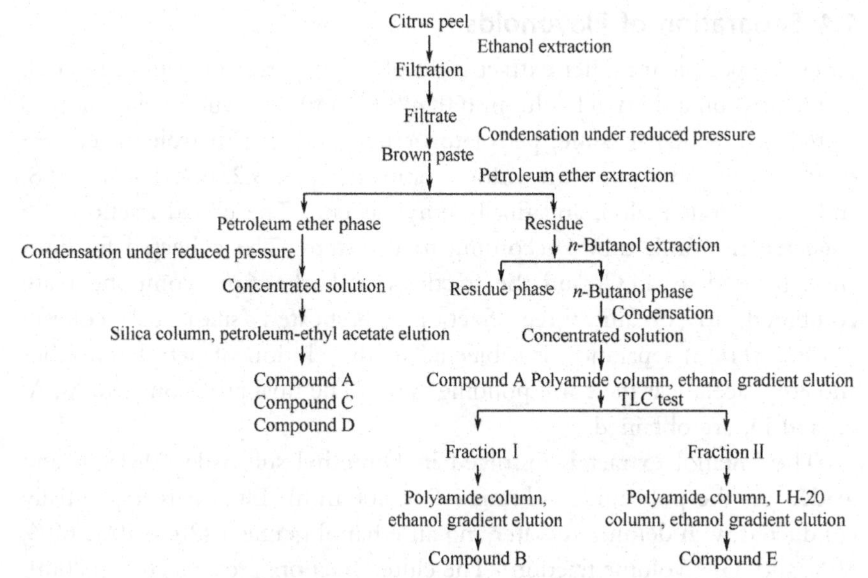

Figure 4.1 Schematic diagram of extraction and purification of flavonoids from citrus peel.

with occasional stirring. After immersion for 24 h, the ethanol extract is filtered and collected through vacuum condensation. The recycled ethanol is adjusted to 95% (v/v) and then transferred to the bucket for another immersion cycle. Immersion is repeated three times. The ethanol extracts are combined and concentrated by rotary evaporation to form a brown extract. The evaporation temperature should be no more than 50°C.

1.3 Ethanol Extraction Treatment

The ethanol extract is degreased with hexane until a colorless hexane layer is formed. The degreased extract is subjected to petroleum ether extraction at three times volume with several repeats until the petroleum ether is colorless. The combined petroleum ether extract is evaporated at an evaporation temperature of less than 40°C. The petroleum ether extract is obtained. After petroleum ether extraction, the residues are subjected to *n*-butanol extraction with three times volume for several repeated extractions until the *n*-butanol phase is colorless. The combined *n*-butanol phase is evaporated to obtain the *n*-butanol extract.

1.4 Separation of Flavonoids

After the petroleum ether extract is dissolved in petroleum ether, filtered, and loaded on a silica gel column (60 × 800 mm). The silica gel column is eluted first with *n*-hexane, petroleum ether, and then petroleum ether–ethyl acetate solution (at the volume ratios of 10:1, 8:2, 6:4, 1:1, 4:6, 2:8, and 1:10, sequentially), and finally ethyl acetate. The eluted fractions are collected in 15mL tubes, according to the steps. The collected fractions are subjected to TLC, and the fractions with the same component are combined to obtain three fractions. Repeated silica gel column (25 × 500 mm) separation is subjected to the elution of petroleum ether and ethyl acetate at the corresponding ratios. The final pure compounds, A, C, and D, are obtained.

The butanol extract is dissolved in Dimethyl sulfoxide (DMSO) and loaded on the polyamide column (60 × 800 mm). Elution is sequentially conducted with deionized water and an ethanol gradient (20%, 40%, 60%, 80%, and 95%, volume fraction). The eluted fractions are collected in 20mL tubes and examined by TLC. The fractions with the same component are combined to obtain two major components. Each fraction mainly containing one component dissolved in DMSO is loaded on the polyamide column or Sephadex LH-20 column (30 × 600 mm) and eluted with an ethanol gradient (5% concentration difference for each gradient). The each fraction is 10 mL and is examined by TLC analysis. The fractions with the same components are combined to obtain two pure compounds, which are compounds B and E.

2. STRUCTURAL IDENTIFICATION OF FLAVONOIDS FROM CITRUS PEEL

2.1 HPLC–MS

The obtained flavonoid monomer is dissolved in methanol and analyzed by liquid chromatography–mass spectrometry (LC-MS).

Liquid phase conditions: chromatography: waters 2690; detector: waters 996; analytical column: Lichrospher C_{18} 2.1 × 250 mm; mobile phase: methanol–water–1% acetic acid gradient elution; column temperature: 30°C; flow rate: 0.3 mL/min; injection volume: 10 μL.

MS conditions: ion mode: EIS^-, EIS^+; capillary voltage: 3.88 kV (EIS^-), 3.87 kV (EIS^+); cone voltage: 30 V (EIS^-), 24 V (EIS^+); ion source temperature: 120°C; off-solvent temperature: 300°C; mass range: 200–800 *m/z*; photomultiplier voltage: 650 V; analyzer vacuum: $2.6e^{-5}$ mBar; gas flow: 4.2 L/h.

2.2 NMR Technique

The isolated flavonoid monomer is dissolved in $CDCl_3$ or DMSO-d_6. ^1H NMR, ^{13}C NMR, and DEPT-NMR are performed on a Bruker NMR spectrometer using tetramethylsilane as the internal standard.

2.3 Identification of the Flavonoid Monomer

Silica gel, polyamide, and Sephadex HL-20 columns are used for the separation and purification of flavonoids from Satsuma mandarin peel. In total, five types of flavonoid monomers are obtained. Structural analysis is performed using the wavelength scanning spectrum, HPLC–MS, and IR, as well as NMR techniques. The five compounds are 3,5,6,7,8,3′,4′-heptamethoxyflavone, 5,6,7,8,4′-pentamethoxyflavone (citrus flavonoids), 3-hydroxy-5,6,7,8,3′,4′-hexamethoxyflavone, 5,6,7,8,3′,4′-hexamethoxyflavone, and hesperidin. Among them, 3-hydroxy-5,6,7,8,3′,4′-hexamethoxyflavone is the first discovered. Structures of compounds A, B, C, D and E are shown in Figs.4.2–4.6.

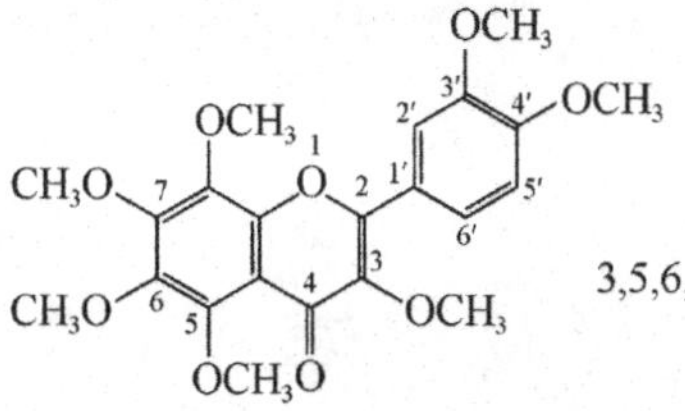

3,5,6,7,8,3′,4′-Heptamethoxyflavone

Figure 4.2 Structure of compound A.

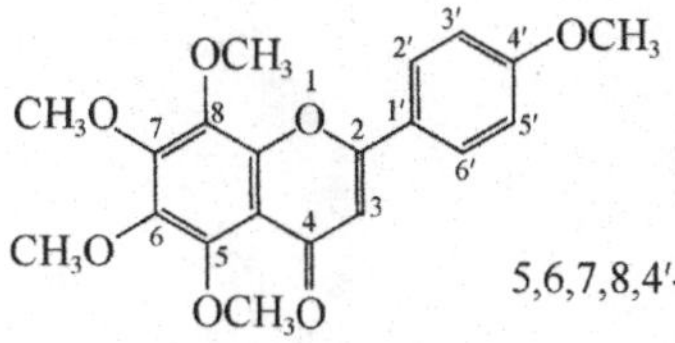

5,6,7,8,4′-Pentamethoxyflavone

Figure 4.3 Structure of compound B.

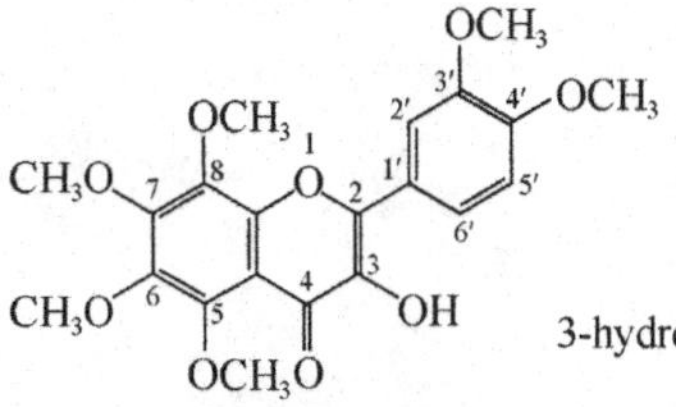

3-hydroxy-5,6,7,8,3′,4′-Hexamethoxyflavone

Figure 4.4 Structure of compound C.

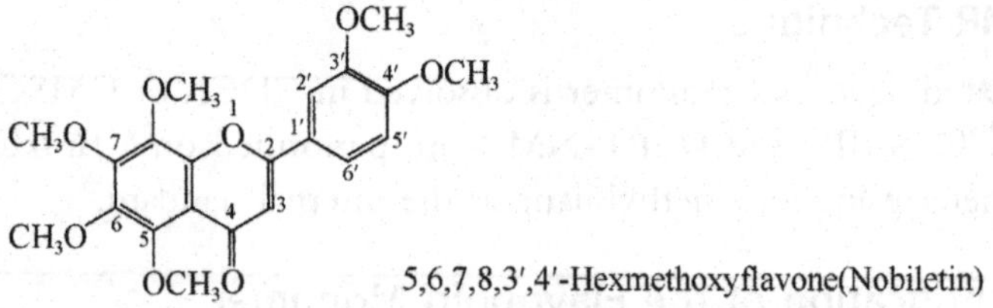

Figure 4.5 Structure of compound D.

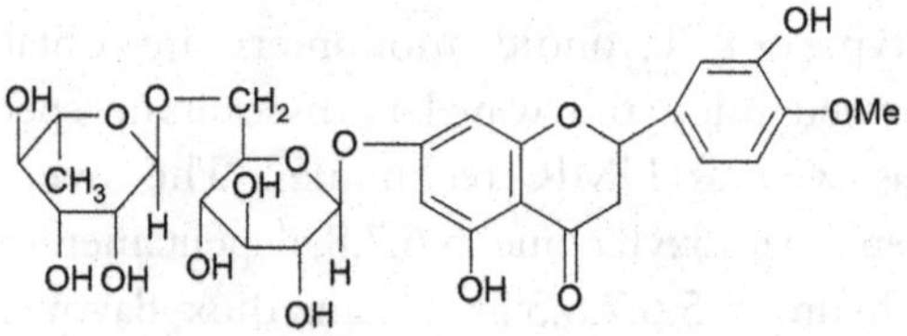

Figure 4.6 Structure of compound E.

CHAPTER 5

Five Types of Semisynthetic Bioactive Flavonoids From Hesperidin

Contents

1. Synthesis Route of Bioactive Flavonoids 66
2. Synthesis of Bioactive Flavonoid Compounds 66
   2.1 Synthesis of Hesperetin 66
   2.2 Synthesis of Persicogenin 67
   2.3 Synthesis of Diosmin 68
   2.4 Synthesis of Diosmetin 68
   2.5 Synthesis of Luteolin 69
   2.6 Synthesis of 7-*O*-Prenyl Hesperetin 69
   2.7 Synthesis of 7-*O*-Farnesyl Hesperetin and 7,3′-*O*, *O*-Bifarnesyl Hesperetin 70
3. Key Points in Flavonoid Synthesis 71
   3.1 Synthesis of Compounds 71
   3.2 Characterization of 7-*O*-Prenyl Hesperetin, 7-*O*-Farnesyl Hesperetin, and 7,3′-*O*, *O*-Bifarnesyl Hesperetin 72

Flavonoids are widely distributed in plants and are a type of bioactive component in medicinal plants, having a broad spectrum of pharmacological activity and low toxicity. Because of the discovery of more flavonoids and pharmacological characteristics, flavonoids have become a research hotspot of natural medicinal exploitation in China. During the rational utilization of biological flavonoid resources, the semisynthetic derivatives or drugs of natural products with high content and physiological bioactivity have the advantages of abundant sources, few synthetic steps, and low production cost, indicating the significance of research and development of these drugs.

Persicogenin (3′,5-dihydroxy-4′,7-dimethoxy flavanone), as a naturally bioactive component isolated from *Prunus persica*, has the function to dilate coronary arteries and reduce myocardial contractility. Luteolin (5,7,3′,4′-hydroxy flavonoids) is an active component isolated from *Dracocephalum* in Xinjiang and Ajuga in Anhui for bronchitis treatment. This natural product has been used in the clinical treatment of cough, as expectorant, and

Comprehensive Utilization of Citrus By-products
ISBN 978-0-12-809785-4
http://dx.doi.org/10.1016/B978-0-12-809785-4.00005-8

anti-inflammation. The clinical application of both medicinal products is limited because of the low extraction efficiency of their flavonoids (approximately 0.02%). In addition, the synthesis of persicogenin and luteolin using phloroglucinol and vanillin as raw materials has the flaws of multiple reaction steps, low yield, and the complex synthetic process. Hesperidin is a natural product present in high amounts in citrus peel and obtained as a by-product during citrus processing. It has a structure similar to that of persicogenin and luteolin. *O*-isopentenylization or *O*-glycosylation can increase the lipophilicity of flavonoids, thereby improving cell membrane permeability and their biological activity in organisms.

Hesperidin can be used as the raw material to synthesize persicogenin and luteolin through hydrolysis of glycosides, selective methylation, dehydrogenation, and demethylation. Meanwhile, hesperidin hydrolysis can result in the generation of hesperetin. Hesperetin can react with prenyl bromide and alkenyl farnesyl bromide to produce 7-*O*-prenyl hesperetin, 7-*O*-farnesyl hesperetin, and 7,3′-*O*, *O*-bifarnesyl hesperetin that have not been reported in literature. The structures of the synthesized products are confirmed by nuclear magnetic resonance (NMR), infrared spectroscopy (IR), and mass spectrum (MS). This synthetic method has the following characteristics: resource abundance of raw materials, simple synthetic process, high yield, and high application value.

1. SYNTHESIS ROUTE OF BIOACTIVE FLAVONOIDS

The synthesis routes of the aforementioned five types of bioactive flavonoids produced using hesperidin as the raw material are shown in Fig. 5.1.

2. SYNTHESIS OF BIOACTIVE FLAVONOID COMPOUNDS

2.1 Synthesis of Hesperetin

Hesperidin (1.50 g, 2.46 mmol) is added to anhydrous methanol (90 mL), and then concentrated sulfuric acid (3 mL) is added to the hesperidin methanol solution with vigorous stirring. The hesperidin methanol solution is heated to reflux for 7.5 h. The reaction mixture is extracted with ethyl acetate (3 × 90 mL). The organic layer is washed with water and 5% sodium carbonate solution to form a colorless layer, and then with saturated sodium chloride solution. It is then dried through anhydrous magnesium sulfate. The ethyl acetate is distilled under reduced pressure to obtain a

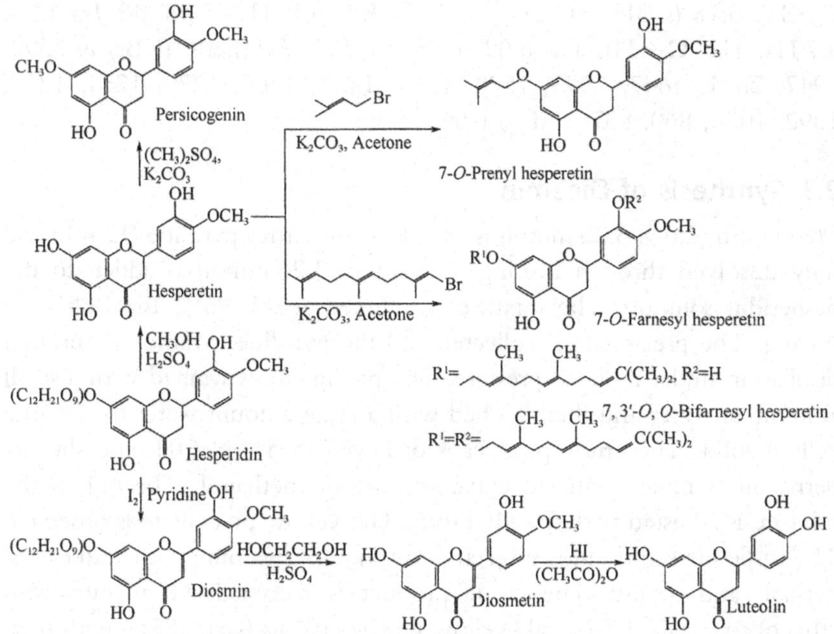

Figure 5.1 Synthesis routes of bioactive flavonoids.

yellow residue. The crude product is recrystallized using ethanol–water (7:3) to obtain 0.56 g pale yellow needle crystals (yield, 82%) with m.p. of 224–226°C: MS (FAB$^-$) *m/z*: 301 (M^+–1); ^{1}H NMR (400 MHz, DMSO-d_6) δ: 2.20 (dd, J = 16.8, 4.8 Hz, 1H, C_3–H), 3.20 (dd, J = 16.8, 12.4 Hz, 1H, C_3–H), 3.77 (s, 3H–OCH_3), and 5.43 (dd, J = 12.4, 4.7 Hz, 1H, C_2-H), 5.88–6.95 (m, 5H, ArH). IR (KBr) ν: 3500, 3114, 3028, 2958, 2744, 1647, 1608, 1504, 1282, 1242, 866, and 816 cm^{-1}.

2.2 Synthesis of Persicogenin

Hesperetin (0.35 g, 1.0 mmol) is added to anhydrous acetone (60 mL) and anhydrous potassium carbonate (0.135 g, 1.0 mmol), and then dimethyl sulfate (0.096 mL, 1.1 mmol) is added to the solution in a dropwise manner during stirring. The hesperetin solution is heated for refluxing for 1 h. Acetone is distilled to obtain a crude product, a yellow powder. The crude product is recrystallized from ethanol to obtain 0.23 g white needle crystals (yield, 73%) with m.p. of 161–162°C: MS (FAB$^+$) *m/z*: 317 (M^++1); ^{1}H NMR (400 MHz, DMSO-d_6) δ: 2.74 (dd, J = 17.2, 4.7 Hz, 1H, C_3–H), 3.26 (dd, J = 17.2, 12.4 Hz, 1H,

C_3−H), 3.78 (s, 3H, −OCH_3), 3.79 (s, 3H,−OCH_3), 5.48 (dd, J = 12.4, 4.7 Hz, 1H, C_2−H), and 6.08−6.95 (m, 5H, ArH). IR (KBr) ν: 3392, 2947, 2841, 1647, 1593, 1570, 1516, 1456, 1365, 1279, 1254, 1230, 1092, 1026, 860, 837, and 814 cm^{-1}.

2.3 Synthesis of Diosmin

Hesperidin (2.0 g, 3.25 mmol) is added to anhydrous pyridine (12 mL) and fully dissolved through heating. I_2 (0.86 g, 3.26 mmol) is added to the hesperidin solution. The mixture is refluxed at 90−95°C for 10 h with stirring. The precipitate is collected and the pyridine is removed through distillation under reduced pressure. The precipitate is washed with a small amount of water and then washed with a large amount water to obtain a yellow solid. The crude product is dissolved in 5% NaOH, and the supernatant is mixed with an equal volume of methanol. The pH of the mixture is adjusted with 5% HCl to 5. The yellow precipitate is observed. The suspended solution is subjected to filtration, washing with water until neutral, and drying. The crude product is recrystallized in anhydrous ethanol to obtain 1.70 g pale yellow needle crystals (yield, 85%) with m.p. of 253−255°C (literature [12] 250−255°C).

2.4 Synthesis of Diosmetin

Diosmin (1.70 g, 2.8 mmol) is added to ethylene glycol (40 mL) and concentrated sulfuric acid (1.7 mL) and then heated in water bath at 90°C or in oil bath at 98°C for 2 h. The resultant yellow clear solution is poured into ice water to precipitate a yellow solid. After filtration under reduced pressure and drying, the dried product is subjected to column chromatography using petroleum ether−ethyl acetate (1:1) as the eluent to obtain 0.69 g yellow powdery crystals (yield, 82%) with m.p. of 254−255°C (literature value [13] 253−255°C). ^{1}H NMR (400 MHz, DMSO-d_6) δ: 3.87 (s, 3H, OCH_3-4′), 6.21 (d, J = 2.0 Hz, 1H, C_6−H), 6.48 (d, J = 2.0 Hz, 1H, C_8−H), 6.77 (s, 1H, C_3−H), 7.10 (d, J = 8.8 Hz, 1H, C_5−H), 7.43 (d, J = 2.0 Hz, 1H, C_2−H), 7.54 (dd, J = 8.8, 2.0 Hz, 1H, C_6'−H), 9.51 (1H, s, ArOH), 10.90 (1H, s, ArOH), and 12.95 (s, 1H, Ar−OH-5). ^{13}C NMR (125 MHz, DMSO-d_6). δ: 163.5 (C-2), 104.1 (C-3), 181.7 (C-4), 157.5 (C-5), 99.1 (C-6), 164.5 (C-7), 94.1 (C-8), 161.8 (C-9), 103.7 (C-10), 118.6 (C-1′), 113.2 (C-2′), 146.7 (C-3′), 151.1 (C-4′), 112.3 (C-5′), 123.1 (C-6′), and 55.8 (OCH_3).

2.5 Synthesis of Luteolin

Diosmetin (1.50 g, 5 mmol) is added to newly distilled deionized water (45 mL) and cooled in a water bath. Then, acetic anhydride (10 mL) is slowly added in a dropwise manner to the diosmetin solution. The mixture is refluxed for 5 h during stirring, cooled, and allowed to stand overnight. The precipitated yellow crystals are washed with 10% sodium bisulfite solution after filtration and then washed with water until neutral. The obtained crude product is dried at low temperature and recrystallized in ethanol to obtain 0.78 g pale yellow needle crystals (yield, 56%) with m.p. of 326–327°C (literature value [6] 326–328°C); MS (FAB^-) m/z: 285 $[M^+-1]$; ^1H NMR (400 MHz, DMSO-d_6) δ: 6.18 (d, J = 2.1 Hz, 1H, C_6–H), 6.42 (d, J = 2.1 Hz, 1H, C_8–H), 6.63 (s, 1H, C_3–H), 6.87 (d, J = 8.2 Hz, 1H, C_5'–H), 7.40–7.44 (m, 2H, C_2'–H, C_6'–H), 9.8 (br, 2H, ArOH), and 12.97 (s, 1H, Ar–OH-5).

2.6 Synthesis of 7-*O*-Prenyl Hesperetin

Hesperetin (102 mg, 0.33 mmol) dissolved in 5 mL of anhydrous acetone is added to a 25-mL single-neck flask and then mixed with anhydrous potassium carbonate (45 mg, 0.33 mmol). The mixture is heated in an oil bath at 50°C with vigorous stirring for 1 h. During heating, prenyl bromide solution (0.06 mL, 0.50 mmol) and 2 mL of acetone are added. After completing the addition for 5 min, TLC is used to monitor the reaction (petroleum ether–ethyl acetate = 2:1) until the point of the starting material disappears. After the reaction is stopped, the reacted mixture is filtered and the residues are rinsed with a small amount of acetone. The crude product after rotary evaporation is loaded on the silica gel chromatography column and subjected to elution with petroleum ether–ethyl acetate (3:1). Approximately 95 mg yellow powder is obtained with (yield, 78%) and m.p. of 78–80°C. MS (FAB^+) m/z: 371 (M^++1). ^1H NMR (400 MHz, $CDCl_3$) δ: 1.72 (s, 3H, C_3''–CH_3), 1.79 (s, 3H, C_3''–CH_3), 2.76 (dd, J = 17.0, 2.8 Hz, 1H, C_3–H), 3.06 (dd, J = 17.0, 11.0 Hz, 1H, C_3–H), 3.90 (s, 3H, –OCH_3), 4.50 (d, J = 6.8 Hz, 2H, C_1''–H), 5.30 (dd, J = 11.0, 2.8 Hz, 1H, C_2–H), 5.43 (t, J = 6.4 Hz, 1H, C_2''–H), 5.84 (d, J = 2.5 Hz, 1H, C_8–H), 6.05 (dd, J = 9.2, 2.5 Hz, 1H, C_6'–H), 6.70–6.93 (m, 2H, C_2'–H, C_5'–H), 7.04 (d, J = 1.7 Hz, 1H, C^6–H), and 12.02 (s, 1H, C_5–OH). IR (KBr) ν: 3433, 2932, 1642, 1573, 1514, 1411, 1370, 1298, 1274, 1198, 1157, 1090, 1020, 868, 805, and 545 cm^{-1}.

2.7 Synthesis of 7-*O*-Farnesyl Hesperetin and 7,3′-*O*,*O*-Bifarnesyl Hesperetin

Hesperetin (102 mg, 0.34 mmol) is placed in a 50 mL round-bottom flask. Then 30 mg (0.22 mmol) of anhydrous K_2CO_3, 7 mL of anhydrous acetone, and 0.09 mL (0.32 mmol) of alkenyl farnesyl bromide are sequentially added at 50°C during stirring for 2 h. If no reaction is observed, 0.2 g (1.4 mmol) of anhydrous K_2CO_3 is added at room temperature and left for 2 h. TLC is used to track the generation of new spots. If new spot is observed, 0.2 g (0.32 mmol) hesperetin and 0.5 g (3.6 mmol) K_2CO_3 are added continuously for 2 h reaction. The reaction is stopped when another new point (may be a disubstituted product) is generated. After, the mixture is loaded on the silica gel column through elution with petroleum ether—ethyl acetate (4:1). A yellow solid, 7-*O*-farnesyl hesperetin, weighing 30 mg is obtained (yield, 20%). The m.p. of 7-*O*-farnesyl hesperetin is 86—88°C. MS (FAB^+) m/z: 507(M^++1), ^1H NMR (400 MHz, $CDCl_3$) δ: 1.50 (s, 6H, $C''-CH_3$, $C_7''-CH_3$), 1.60 (s, 3H, $C_{11}''-CH_3$), 1.70 (s, 3H, $C_3''-CH_3$), 1.98—2.10 (m, 8H, $C_4''-CH_2$, $C_5''-CH_2$, $C_8''-CH_2$, $C_9''-CH_2$), 2.80 (dd, $J = 16.9$, 4.8 Hz, 1H, C_3-H cis), 3.10 (dd, $J = 17.0$, 11.2 Hz, 1H, C_3-H trans), 3.90 (s, 3H, $C_4'-OCH_3$), 4.54 (d, 2H, $C_1''-CH_2$), 5.10 (d, 2H, $C_6''-H$, $C_{10}''-H$), 5.30 (dd, $J = 11.1$, 4.8 Hz, 1H, C_2-H), 5.43 (t, 1H, $C_2''-H$), 5.83 (s, 1H, $C_3'-OH$), 6.03 (d, $J = 2.5$ Hz, 1H, C_6-H), 6.07 (d, $J = 2.5$ Hz, 1H, C_8-H), 6.85—6.93 (m, 2H, $C_2'-H$, $C_5'-H$), 7.04 (d, $J = 8$ Hz, 1H, $C_6'-H$), and 12.02 (s, 1H, C_5-OH). ^{13}C NMR (125 MHz, $CDCl_3$) δ: 195.8, 167.4, 164.1, 162.8, 146.9, 145.9, 142.2, 135.6, 131.6, 131.3, 124.3, 123.5, 118.4, 118.1, 112.7, 110.7, 103.1, 95.7, 94.8, 78.9, 65.4, 56.1, 43.2, 39.7, 39.5, 26.7, 26.2, 25.7, 17.7, 16.7, and 16.0. A yellow liquid, 7,3′-*O*,*O*-bifarnesyl hesperetin, weighing 20 mg is obtained (yield, 18%). MS (FAB^+) m/z: 711 (M^++1), ^1H NMR (400 MHz, $CDCl_3$), δ: 1.59 (s, 12H, $C_{11}''-CH_3$, $C_7''-CH_3$, $C_{11}'''-CH_3$, $C_7'''-CH_3$), 1.67 (s, 6H, $C_{11}''-CH_3$, $C_{11}'''-CH_3$), 1.72 (d, 6H, $C_3''-CH_3$, $C_3'''-CH_3$), 1.93—2.13 (m, 16H, $C_4''-CH_2$, $C_5''-CH_2$, $C_8''-CH_2$, $C_9''-CH_2$, C_4'''-CH_2, C_5'''-CH_2, C_8'''-CH_2, C_9'''-CH_2), 2.80 (dd, $J = 17.0$, 4.9 Hz, 1H, C_3-H cis), 3.11 (dd, $J = 17.0$, 11.0 Hz,1H, C_3-H trans), 3.88 (s, 3H, $C_4'-OCH_3$), 4.54 (d, 2H, $C_1''-CH_2$), 4.64 (d, 2H, $C_1'''-CH_2$), 5.09 (m, 4H, $C_6''-H$, $C_{10}''-H$, $C_6'''-H$, $C_{10}'''-H$), 5.30 (dd, $J = 11.0$, 4.9 Hz, 1H, C_2-H), 5.44 (t, 1H, $C_2''-H$), 5.53 (t, 1H, $C_2'''-H$), 6.03 (d, $J = 2.5$ Hz, 1H, C_6-H), 6.06 (d, $J = 2.5$ Hz, 1H, C_8-H), 6.90 (d, 1H, $C_6'-H$), 6.96—6.99 (m, 2H, $C_2'-H$, $C_5'-H$), and 12.01 (s, 1H, C_5-OH).

3. KEY POINTS IN FLAVONOID SYNTHESIS

3.1 Synthesis of Compounds

During the selective methylation of hesperetin, on the basis of the activity of hydroxyl groups at different positions in flavanones, in general, $C_7{-}OH > C_3{-}OH > C_5{-}OH$, anhydrous potassium carbonate can be used as a condensing agent and dimethyl sulfate as a methylation reagent. During this reaction, when the molar ratio of hesperetin and anhydrous potassium carbonate is less than 1:1, recrystallization results in two types of crystals including yellow crystals (by-products) and white crystals (final products). When the ratio is to 1:1, recrystallization can result in only white crystals without yellow ones, which may be due to the structure of flavanones. Although phenolic hydroxyl groups at different positions reveal different activities, potassium carbonate can react with the phenolic hydroxyl group to generate potassium salt. Because dimethyl sulfate is a good methylation reagent, more than one product is generated with an overdose of potassium carbonate, and the yields are therefore affected. Furthermore, flavanones in alkaline solution can result in the breakage of the pyridine ring in molecules and generate yellow chalcone. After acidification, it can be cyclized into the original flavanone, as shown in Fig. 5.2.

Therefore, it is critical to control the amount of anhydrous potassium carbonate in the selective methylation reaction. The mass ratio of hesperetin and anhydrous potassium carbonate at 1:1 is a better choice.

In diosmin synthesis, it is safer to use I_2/pyr instead of NBS/benzoyl peroxide during dehydrogenation. Free hydroxyl groups are not necessary for the protection of acetylation. At the end of the reaction, because of the different solubilities of hesperidin and diosmin in pyridine, pyridine can be removed through distillation under reduced pressure. The precipitated product is filtered and washed with water to obtain diosmin at a high yield.

Diosmetin is synthesized through the hydrolysis of glycosidic bonds in the presence of ethylene glycol and H_2SO_4. The application of the CH_3OH/H_2SO_4 or CH_3OH/HCl mixture in previous studies cannot produce ideal results. In the presence of sulfuric acid, flavonoids can be

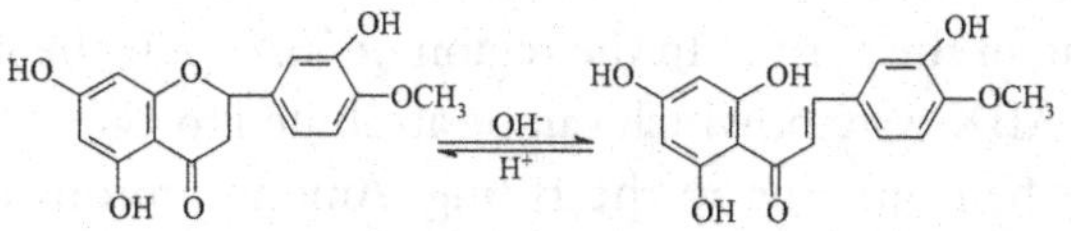

Figure 5.2 Acidification reduction of flavanone.

easily oxidized and result in low yield under the condition of high temperature. In contrast, the reaction cannot be completed under the low temperature condition. The experimentally determined temperature is 90–98°C. Based on monitoring by TLC, the optimal reaction time is 2 h. After the reaction, the reaction mixture can be directly poured into ice water to precipitate flavonoid aglycone. This method has a simple operation and produces a high yield.

Hydriodic acid is slowly added to acetic anhydride is used for demethylation. After the reaction is stopped, the mixture is cooled and allowed to stand to ensure the precipitation of the product as soon as possible. The filtered crude product can be recrystallized in ethanol to obtain luteolin.

During the synthesis of 7-*O*-prenyl hesperetin, 7-*O*-farnesyl hesperetin, and 7,3′-*O*, *O*-bifarnesyl hesperetin, the activities of phenolic hydroxyl groups at different positions in flavanones are different, in general, C_7–OH > C_3–OH > C_5–OH. Therefore, during selective isopentenylization and farnesylization, the amount of potassium carbonate is critical. The optimal molar ratio of hesperetin and anhydrous potassium carbonate should be controlled at 1:1. Prenyl bromide and farnesyl bromide easily decompose at a high temperature. After repeated trial and error, the optimal temperature of the reaction is determined to be 50°C. On the other hand, it is better to generate a large amount of oxygen ions and then add prenyl bromide or farnesyl bromide, which can result in the immediate reaction and complete the reaction in a short time with a high yield. In case of overdose of farnesyl bromide, the reaction will generate 7,3′-*O*, *O*-bifarnesyl hesperetin.

3.2 Characterization of 7-*O*-Prenyl Hesperetin, 7-*O*-Farnesyl Hesperetin, and 7,3′-*O*, *O*-Bifarnesyl Hesperetin

In FAB^+-MS (MNBA as the matrix) of 7-*O*-prenyl hesperetin, 7-*O*-farnesyl hesperetin, and 7,3′-*O*, *O*-bifarnesyl hesperetin, m/z are 371 (M^++1), 507 (M^++1), and 711 (M^++1), respectively, which confirmed that their molecular weights are 370, 506, and 710, respectively.

In ^{1}H NMR spectra of 7-*O*-prenyl hesperetin, 7-*O*-farnesyl hesperetin, and 7,3′-*O*, O-bifarnesyl hesperetin, the doublets ($J = 2.5$ Hz) are observed in δ5.9–6.1 and 6.0–6.4 regions, which come from C_6–H and C_8–H of the phenyl ring in the A ring. In the region of δ6.7–7.1, the multiple peaks belong to the ABX system, which can be attributed to C_2'–H, C_5'–H, and C_6'–H of the benzene ring in the B ring. Among protons in the C ring, coupling between C_2–H and two C_3–H ($J_{trans} = 11.0$ Hz, $J_{cis} = 4.8$ Hz) is observed so that a doublet is shown up at the central location of δ5.3.

Geminal coupling of the two C_3—H protons ($J = 17.0$ Hz) and its coupling with H-2 at the neighborhood location leads to two respective doublet peaks observed at the central location of $\delta 2.9$. The single peak around $\delta 3.9$ should be the hydrogen in the methoxyl group from the B ring.

The chemical shifts of protons in the O-isopentenyl group and the farnesyl alkenyl group should be in the range of $\delta 1.5-5.6$. Wherein, the chemical shifts of protons in the two methoxyl groups in isopentenyl and the four methyl groups in farnesyl should be in the range of $\delta 1.58-1.66$; the chemical shifts of four methylenes without the connection to oxygen should be in the range of $\delta 1.98-2.02$; and the chemical shifts of the methylene with the connection to oxygen should be in the range of $\delta 4.5-4.54$, indicating doublets. The chemical shifts of the alkenyl group in the isoprenyl and three alkenyl groups should be in the range of $\delta 5.11-5.42$.

Furthermore, the ^{13}C NMR spectrum of 7-*O*-farnesyl hesperetin shows 31 peaks in the range of $\delta 16.0-195.8$, which confirms that the compound has 31 carbons. The number of carbon and chemical shift are consistent with its chemical structure. In the IR spectrum of 7-*O*-isoprenyl hesperetin, because of the formation of hydrogen bonds between carbonyl in flavonoids and C_5—OH in the A ring, characteristic absorption of the carbonyl group reveals the shift to a lower wave number and is observed at 1642 cm^{-1}. The characteristic absorption of hydroxyl groups is observed at approximately 3433 cm^{-1}.

CHAPTER 6

Drying of Citrus Peel and Processing of Foods and Feeds

Contents

1. Drying of Citrus Peel 76
  1.1 Technological Process 76
  1.2 Key Operating Points 76
    1.2.1 Citrus Peel as Raw Material 76
    1.2.2 Removal of Pesticides Through Soaking 76
    1.2.3 Rinsing 77
    1.2.4 Segmentation 77
    1.2.5 Drying 77
    1.2.6 Mechanical Screening 77
    1.2.7 Manual Sorting 77
    1.2.8 Wind and Magnetic Sorting 77
    1.2.9 Packaging 77
    1.2.10 Storage 77
  1.3 Quality Requirements 77
2. Processed Foods of Citrus Peel 78
  2.1 Processing Technology of Health-Promoting Citrus Peel Sauce 78
    2.1.1 Technological Process 78
    2.1.2 Key Operating Points 79
    2.1.3 Quality Index 79
  2.2 Processing of Candies With Citrus Peel 80
    2.2.1 Technological Process 80
    2.2.2 Key Operating Points 80
    2.2.3 Product Quality 80
3. Production of Various Citrus Feed Additives 80
  3.1 Feeds of Dried Citrus Peel 81
    3.1.1 Technological Process 81
    3.1.2 Pros and Cons 81
    3.1.3 Evaluation of Nutrient Scores 81
  3.2 Feeds of Citrus Peel Powder 82
    3.2.1 Technological Process 82
    3.2.2 Pros and Cons 82
    3.2.3 Evaluation of Nutrients 82
    3.2.4 Evaluation of Animal Feeds 82

Comprehensive Utilization of Citrus By-products
ISBN 978-0-12-809785-4
http://dx.doi.org/10.1016/B978-0-12-809785-4.00006-X

3.3 Citrus Pulp Silage 82
3.3.1 Technological Process 82
3.3.2 Pros and Cons 83
3.3.3 Evaluation of Nutrients 83
3.3.4 Evaluation of Animal Feeding 83

1. DRYING OF CITRUS PEEL

Because of the intense attention on food cellulose, reprocessing of citrus peel is becoming increasingly important. In recent years, the amount of processed and exported dried citrus peel in the major producing areas in China reveals a rising trend. The major exporting countries include Japan, Southeast Asian countries, and other Asian countries. The export amount of citrus peel in recent years is at almost 1000 tons and the price is $600–1000/ton, indicating better economic benefit. Citrus peel is in great demand in traditional Chinese medicine, however, the quality of those citrus peel that is randomly collected cannot be ensured.

1.1 Technological Process

Citrus peel as the raw material ⟶ Washing and soaking ⟶ Rinsing ⟶ Cutting, and screening ⟶ Drying ⟶ Mechanical screening ⟶ Manual screening ⟶ Winding and magnetic screening ⟶ Packaging ⟶ Storage.

1.2 Key Operating Points

1.2.1 Citrus Peel as Raw Material

The residues of Satsuma mandarin peel obtained after orange syrup processing are chosen as the raw materials. In general, the newly stripped peel should be treated within 2 h, and water content should be controlled at approximately 82%. Meanwhile, no mildew, no odor, no color change, and no softening are observed. In short, timely treatments are required.

1.2.2 Removal of Pesticides Through Soaking

Considering surfactant as the major component to prepare pesticide-removing agents, 1.5% pesticide-removing agent is used to soak the samples for 30 min. During soaking, stirring is not necessary and soaking duration cannot be too long. In addition, soaking should be conducted in stainless containers.

1.2.3 Rinsing
The pesticide remover is cleaned with fresh water through soaking for 10 min. After drainage, the flowing water is used to rinse the samples to pH 7.0. The citrus peel is placed in a plastic box with holes to remove the water.

1.2.4 Segmentation
According to different requirements, the citrus peel is cut into different blocks and filaments.

1.2.5 Drying
On a belt dryer, the citrus peel is dried by passing through hot air at 150°C with a wind speed of 2 m/s. The emission speed of waste vapor is controlled at 8000 m^3/h. The citrus peel is kept on the dryer for 50 min and released when the peel suffers a weight loss of 65%. The released citrus peel is naturally cooled to 30°C. During the cooling process, the weight loss of the citrus peel is 5%. Finally, the citrus peel is dried with hot air at 85°C for 30 min so that the water content can reach the standard.

1.2.6 Mechanical Screening
The debris and powder in the dried citrus peel are removed through mechanical screening.

1.2.7 Manual Sorting
The discolored and obviously coked citrus peel is sorted manually. Meanwhile, the incomplete debris and various foreign bodies are removed.

1.2.8 Wind and Magnetic Sorting
The hair and metal foreign materials in the dried citrus peel are removed using a wind sorting machine with a powerful wind and magnetic force.

1.2.9 Packaging
The sealed plastic bags are used as inner packaging.

1.2.10 Storage
The dried citrus peel is stored in a drying room to prevent contact with moisture.

1.3 Quality Requirements

1. The color of citrus peel should be orange yellow or a little darker. The yellow skin layer has an inherent shiny color of various citrus varieties.

The white skin layer is almost white or light yellow and has an inherent flavor of citrus varieties without odor. The citrus peel allows for the presence of a very small amount of powder, and the presence of any foreign material is forbidden.

2. Water content of less than 10%, arsenic content (As) of less than or equal to 0.5 mg/kg, lead (Pb) of less than or equal to 1.0 mg/kg, copper (Cu) of less than or equal to 10 mg/kg, and ash of less than or equal to 4%.
3. Total number of colonies is less than 1000 cfu/g, mold count is less than 50 cfu/g, and *Escherichia coli* and pathological bacteria cannot be detected.
4. The limit of pesticide residues is in compliance with relevant national regulations.

2. PROCESSED FOODS OF CITRUS PEEL

2.1 Processing Technology of Health-Promoting Citrus Peel Sauce

With the rapid development of the catering industry and continuous improvement in people's standard of living, the production and development of condiments have gained unprecedented prosperity and people have attached greater importance to condiments. Condiments not only of good color, fragrance, and taste but also containing more nutrition, health, and convenience are highly demanded. Therefore, a single spice cannot fully meet the demands of people. Citrus fruits are a common favorite. According to the theory of Chinese medicine, citrus peel can treat cough and chest tightness through rectifying qi and removing phlegm. Citrus peel can be used as the raw material to develop a complex sauce with a typical flavor of the peel. This sauce is a type of health-promoting condiment formed through the complex fermentation of processed and treated citrus peel on the basis of the broad bean sauce. The development of this functional citrus peel sauce cannot only increase sauce varieties but also provide an effective strategy for the comprehensive utilization of citrus peel.

2.1.1 Technological Process

Citrus peel ⟶ Washing ⟶ Filaments ⟶ Adding salt solution ⟶ Cooking ⟶ Soaking ⟶ Condensation ⟶ Broad bean Stripping peel ⟶ Soaking ⟶ Steaming materials ⟶ Inoculation for preparing fermentation starter ⟶ Fermentation ⟶ Maturation ⟶ Sterilizing and packaging ⟶ Final product.

2.1.2 Key Operating Points

1. Cleaning of citrus peel: Cleaning of citrus peel is to remove the dust, microorganisms, pesticide residues, and rind or the rot peel to ensure product quality.
2. Filaments of citrus peel: The cleaned citrus peel is cut into filaments with a length of 30 mm and width of 1–3 mm.
3. Boiling, soaking, and concentrating: The citrus peel is boiled in 20% NaCl solution and soaked in brine after boiling. The extract of citrus peel is concentrated in a mezzanine pot to 60% soluble solids for future use.
4. Swelling of peeled broad beans: The broad beans are peeled, cleaned, and placed in fresh water for soaking until the section has no white core, the bean volume is increased by 2–2.5-fold, and the weight of the beans is increased by two fold.
5. Steaming raw materials and preparing koji: The broad beans are placed in a steamer for steaming. Flour is added to the beans at a proportion of 30%, and 0.3% koji is added to the mixture. Then, the mixture is moved into an incubator for culturing. Ventilation koji time is approximately 48 h.
6. Paste fermentation: The starter blocks are moved to a fermentation tank and mixed with the citrus peel extract at a ratio of 5:1. Then, 4 kg vegetable oil, 1 kg spices, and 4 kg wine are added to the 100 kg mixture in the fermentation tank. The samples in the fermentation tank are compacted. When the temperature is increased to 40°C, 140 kg of the brine preheated to 65°C 15Bx is sprayed on the 100 kg of sauce paste and covered with a layer of salt to seal the fermentation tank. When the temperature of the fermentation tank is increased to 45°C, fermentation is conducted for 10 days until the maturation of the sauce paste. After fermentation and maturation, 8 kg salt and 10 kg water are added to the 100 kg matured sauce paste. After mixing well, continuous fermentation at 45°C for another 4–5 days is required.
7. Sterilization and packaging: The final sauce paste products are heated to 120°C for 20-min sterilization and packaged.

2.1.3 Quality Index

The brown sauce has red-brown filaments with an even distribution. The final sauce product is bright and shiny, rich in sauce fragrance, delicious, and rich in a distinctive citrus peel flavor.

2.2 Processing of Candies With Citrus Peel

Citrus peel is used as a raw material to prepare candy that is a type of food with optimal color, fragrance, and taste and has medicinal effects.

2.2.1 Technological Process

Citrus peel → Filaments → Soaking → Rinsing → Presteaming → Cooking with sugar → Drying with natural drainage → Baking → Final product.

2.2.2 Key Operating Points

1. Selection of raw materials: Citrus peel without rotten parts is selected and cleaned with fresh water (dry citrus peel needs to be soaked in water for 2–3 h).
2. Filaments: The citrus peel is cut into filaments with a length of 5 cm and width of 0.5 cm.
3. Soaking: The chopped citrus peel is soaked in 1% lime to remove the odor.
4. Rinsing: The citrus peel with debittering is rinsed with fresh water several times to remove the astringency of residual lime.
5. Precooking: The treated citrus peel strips are placed in boiling water and cooked for several minutes. They should be fully cooked.
6. Cooking with sugar: A certain amount of white sugar is added to the treated citrus peel, and the peel is cooked for 5 min. Then, the sugar-cooked citrus peel is soaked in sugar solution for 24 h. During the first cooking with sugar, the content of sugar is 45%, and it will be increased by 10–15% during later cycles of cooking until the content of sugar is up to 75%.
7. Drainage: The citrus peel is collected from the last cooking with sugar and dried through natural drainage.
8. Baking: The citrus peel is placed in the oven or the baking room with a temperature of 60–70°C until it is 70–80% dry. After cooling, the final products are achieved.

2.2.3 Product Quality

Citrus peel candies have a golden color and are transparent. They have a soft texture and are therefore suitable for people of all ages.

3. PRODUCTION OF VARIOUS CITRUS FEED ADDITIVES

Citrus peel is rich in nutrients such as carbohydrates, fats, and vitamins, which are important to various animals. Its nutrients are compared with those in corn and rice, as shown in Table 6.1.

Table 6.1 Nutrients in citrus peel, corn, and rice

Index	Citrus peel	Corn	Rice
Dry materials%	92.23	88.00	87.00
Crude protein%	6.25	8.50	6.80
Crude fat%	4.40	4.30	2.50
Crude fiber%	16.27	1.60	8.20
Nonnitrogen extract%	61.51	72.20	56.60
Ash%	3.80	1.70	4.50
Cow energy unit (NND/kg)	2.48	2.87	2.19
Digestive energy (Mcal/kg)	3.38	3.25	2.95
Metabolism energy (Mcal/kg)	2.78	2.64	2.40
Calcium%	0.34	0.02	0.01
Phosphorus%	0.25	0.21	0.27
Fe (mg/kg)	204.8	16.0	15.0
Zn (mg/kg)	117.1	18.7	17.2
Vitamin B_1(mg/100 g)	2.00	0.34	0.51
Vitamin B_2(mg/100 g)	0.36	0.10	0.07
Vitamin B_6(mg/100 g)	0.19	—	—
Vitamin E (mg/100 g)	5.03	2.20	—
Vitamin C (mg/100 g)	28.8	0	0
Amino acid %	3.81	4.10	3.20

3.1 Feeds of Dried Citrus Peel

3.1.1 Technological Process

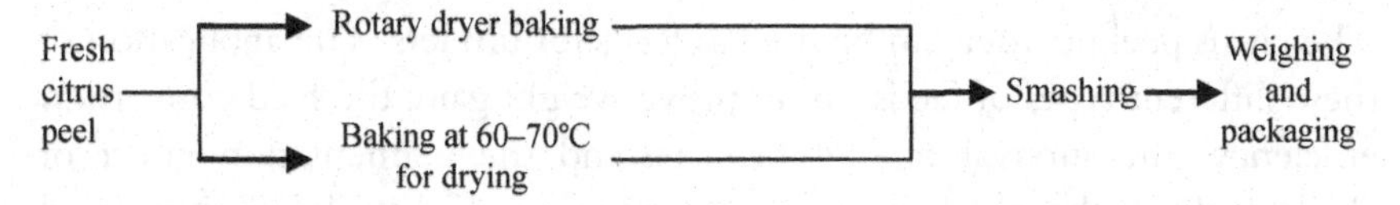

3.1.2 Pros and Cons

This method is simple and fast, and the final products can be easily stored and transported, but the method needs higher energy costs. If it can use local conditions to apply waste heat for drying, this method should be a good approach.

3.1.3 Evaluation of Nutrient Scores

After sampling measurement, dried citrus residue (Satsuma mandarin) feeds contain 9.8% crude protein (CP), 2.2% crude fat (EE), 69.4% nitrogen-free extract (NFE), 12.56% crude fiber (CF), 1.039% calcium (Ca), and 0.10% phosphorus (P) and have better palatability. The results show that dried citrus residues are rich in nutrients and are suitable as a source of high-quality forage for ruminants such as cattle and sheep that can be supplied throughout year.

3.2 Feeds of Citrus Peel Powder

3.2.1 Technological Process

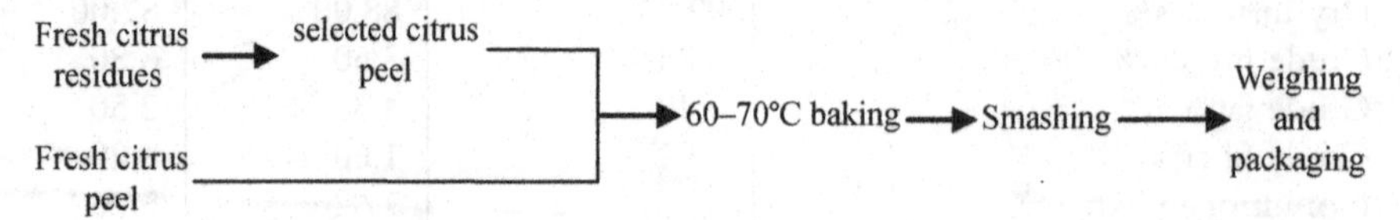

3.2.2 Pros and Cons

This method uses the thin rind and refined citrus peel as the major raw material. The pros and cons of this method are similar to that used with the dried citrus residues.

3.2.3 Evaluation of Nutrients

After sampling measurement, citrus peel powder contains 8.17% CP, 2.60% EE, 61.40% NFE, 14.32% CF, 0.60% Ca, and 0.07% P. The contents of lysine, methionine, and threonine are 0.13%, 0.05%, and 0.15%, respectively. The citrus peel has an orange flavor and a good palatability. Orange is rich in nutrients, especially in carotenoids. Its characteristics are suitable for use as multifunctional feed additives for monogastric animals such as pig and poultry, which can improve flavor and pigmentation in the skin and egg yolk, as well as improve resistance against bacteria.

3.2.4 Evaluation of Animal Feeds

The citrus peel powder can be used as feeds for broilers. The application of these different types of feeds can improve weight gain, the feed conversion efficiency, the survival rate of broilers, and the pigmentation effect of chicken. In Yisha chicken feeds, the addition of 1.0–1.5% citrus peel powder has an excellent egg yolk pigmentation-promoting effect.

3.3 Citrus Pulp Silage

3.3.1 Technological Process

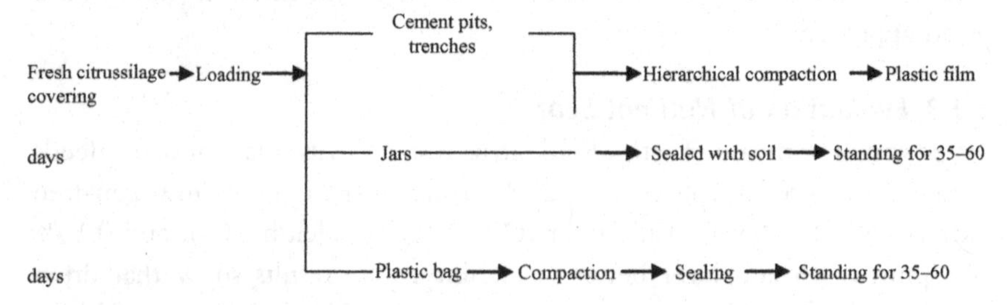

3.3.2 Pros and Cons

This method is fast and economical. A large amount of processed fresh citrus pulp can be preserved well. In addition, it can be stored in a mixture form according to demand. Citrus pulp silage is the most simple and cost-effective, high-quality preservation approach, which is worthy of further exploration and vigorous promotion.

3.3.3 Evaluation of Nutrients

Citrus pulp silage can keep the original color and nutrients of fresh slag for a long term. Fresh Satsuma mandarin pomace contains 1.77% CP, 0.68% EE, 14.5% NFE, 2.24% CF, 0.16% Ca, 0.05% P, and 77.4% H_2O. In addition, this type of juicy feed is rich in vitamins and minerals and has a strong fruit acid fragrance and palatability. It is suitable for ruminant animals; especially, it can be the supplementary feed of high-quality green forage for cows. When the acidity of the feed is too high, appropriate sodium bicarbonate can be added as the buffer.

3.3.4 Evaluation of Animal Feeding

Citrus pulp and dried citrus pomace are supplementarily added to the daily diets of cows. They are favored by cows and have excellent economic benefits. Under this experimental condition, the production of milk can increase by 8.0%, and the daily income per cow can increase by 2.40 yuan.

CHAPTER 7

Biotransformation of Citrus Peel

Contents

1. Production of Food 85
   1.1 Citrus Vinegar 85
   1.2 Lactic Acid Drinks 86
   1.3 Natural Turbidity Agent 86
2. Production of High-Protein Feeds 86
3. Production of Ethanol, Mushroom Cultivation, and Preparation of Other Biological Products 87
   3.1 Production of Ethanol 87
   3.2 Cultivation of Mushrooms 88
   3.3 Other Biotransformation Products of Citrus Peel 88
      3.3.1 Preparation of Citrus Peel Plastic Products 88
      3.3.2 Production of Biogas 89
      3.3.3 Preparation of Citric Acid 89
      3.3.4 Preparation of Activated Carbon 89

In addition to the extraction of functional components and the processing of foods and feeds from citrus peel, citrus peel can be transformed by microbes to produce vinegar, alcohol, high-protein animal feeds, and media for the cultivation of edible fungi.

1. PRODUCTION OF FOOD

1.1 Citrus Vinegar

With the improvement in people's standard of living, requirements with regard to the nutrition, flavor, and color of various beverages have increased. Healthy juice drinks have attracted increasing consumer attention. Because citrus peel is not only rich in glucose, fructose, sucrose, and amino acids but also contains a certain amount of thiamine, riboflavin, vitamins, bioflavonoids, carotenoids, and other physiologically active substances, it is used as a raw material for beverage production.

The brewing process of citrus vinegar is as follows:

Citrus peel ⟶ Cleaning ⟶ Juicing ⟶ Filtration ⟶ Adding sugar ⟶ Tuning Brix ⟶ Sterilizing ⟶ Cooling ⟶ Adding yeast

Comprehensive Utilization of Citrus By-products
ISBN 978-0-12-809785-4
http://dx.doi.org/10.1016/B978-0-12-809785-4.00007-1

⟶ Alcohol fermentation ⟶ Clarifying ⟶ Fruit wine ⟶ Sterilizing ⟶ Adding acetic acid bacteria ⟶ Filtration ⟶ Sterilizing ⟶ Cooling ⟶ Final product.

1.2 Lactic Acid Drinks

Yongxian Wu, Peirong Yang, Wangcheng Gu, and Casimir et al. have attempted to inoculate *Lactobacilli* in citrus to prepare lactic acid drinks or add juice pulp and crushed citrus peel to the juice according to different demands, thereby producing fruit drinks of superior quality.

1.3 Natural Turbidity Agent

In addition, because waste materials obtained by citrus processing contain large amount of pigments and various polysaccharides, it can be used as the raw material for producing a natural turbidity agent for use in the soft drink industry. Lashkajani et al. used lemon, orange, or grapefruit peel as the raw material to prepare beverage-clouding agents with excellent appearance and stability; this is a patented process.

2. PRODUCTION OF HIGH-PROTEIN FEEDS

After processing, citrus peel is rich in abundant nutrients such as carbohydrates, fats, vitamins, amino acids, and minerals. In Brazil and the United States, 40% of the dried citrus peel has been prepared into pellet feeds each year. However, citrus pulp has very bitter and low protein content and is poorly absorbed by animals. Chemical and microbiological methods can improve the feeding effect. Therefore, Houjiu Wu et al. applied microbiological fermentation of citrus peel to produce high-protein feeds. They used citrus peel (including seeds and capsule clothing) obtained after juicing using an FMC citrus juicer as the fermentation substrate to perform fermentation. *Aspergillus niger* II and *Candida utilis* I are the strains used for this process. Fermentation method: The abovementioned bacteria ferments the substrate at room temperature. First, two types of strains are cultured for amplification. The strains at the dose of 1% dried material weight are inoculated to the sterilized citrus peel, which are cooled to 40°C. After mixing well, the fermentation tanks are sealed with a plastic film to prevent bacterial infection. After the surface of citrus peel is covered with mycelium, the fermented pomace is discharged and dried at 60–80°C to water content below 12%, and then ground into powder and transferred to a plastic bag for storage.

Technological process:

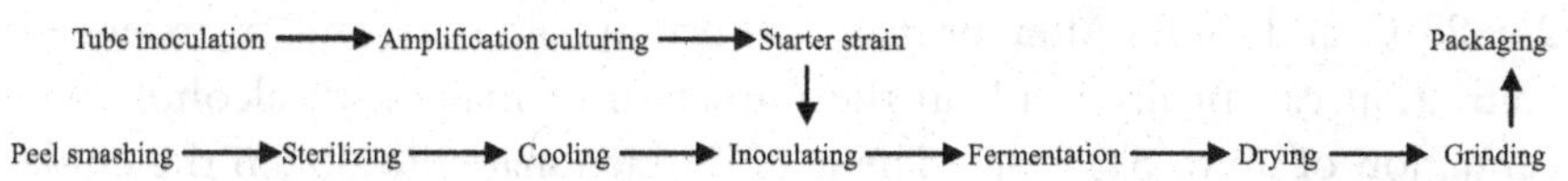

Fermented citrus peel has a high nutritional value. Compared with unfermented citrus peel, fermented citrus peel reveals the enhancement in crude protein, amino acids, crude fat, and vitamins by 50%, 55%, 24%, and 10–60%, respectively, as well as the reduction in crude dietary fiber by 17%, suggesting that citrus peel subjected to fermentation not only greatly improves the contents of protein and amino acids and makes up for the original shortage but also promotes the further increase in vitamins. On the other hand, high calcium and phosphorus contents result in a higher nutrition level of fermented citrus peel feeds. Compared with fine feed corn, citrus peel–fermented feed reveals the increase in crude protein by 14%, amino acids by 46%, crude fat by 25%, calcium by 40-fold, phosphorus by 9-fold, iron by 26-fold, zinc by 5-fold, and vitamins by 3–20-fold, which is also better than rice. Therefore, citrus peel–fermented feed can be used as the feed additive with plentiful mineral nutrients and vitamins.

Cows are continuously fed fermented citrus peel feeds for 35 days. The results show that daily feeding of 0.75 kg fermented citrus peel fed to each cow is best for milk production, and daily milk production by each cow is increased by 1.28 kg. In addition, the milk-promoting effect is obvious, and the milk-promoting benefit can be realized over one week after the second day after the feeding of fermented citrus peel feeds.

3. PRODUCTION OF ETHANOL, MUSHROOM CULTIVATION, AND PREPARATION OF OTHER BIOLOGICAL PRODUCTS

3.1 Production of Ethanol

Pomace processed by the juicer is minced and mixed, and then the lime is added with shaking. The mixture is placed into the squeezing machine for 30 min of mixing. The sugar solution with 13°Brix is achieved, and then pH is adjusted to 5 using 6 mol/L H_2SO_4. The solution is filtered for future use. The better fermentation strain for citrus sugar is screened from yeast strains. The slant medium is composed of 13°Brix wort coupled with 1.5%

agar, and fermentation medium is composed of citrus peel sugar. The sterilization temperature, pressure, and duration are 110–120°C, 0.1 MPa, and 20 min, respectively. Fermentation temperature and duration are 26–28°C and 24 h. After primary cultivation, 48 h secondary extended cultivation can finally result in the formation of mash with alcohol concentration of up to 5% (v/v). The mash is fractionated to obtain the crude fraction of ethanol. The CaO is added to the crude fraction of the ethanol, and distillation is conducted to obtain the refined ethanol product.

3.2 Cultivation of Mushrooms

Edible mushrooms such as *Pleurotus ostreatus*, Enoki mushroom, and *Ganoderma lucidum*, are a type of health food with high protein, low fat, low calories, comprehensive nutrients, and a unique flavor. Cottonseed hulls and sawdust are currently used as cultivation materials. Artificial cultivation and production of mushrooms under natural conditions can be greatly affected by the supply of raw materials and seasonal changes, making it difficult to meet market demands. The research results show that the carbon–nitrogen ratio of pomace is suitable for the growth of mushrooms, and it shows fast mycelium growth and high mushroom production. Therefore, pomace is a good material for mushroom production. For mushroom cultivation, dried pomace powder can be up to 70% of the total materials supplemented with 2% lime, wheat bran, and sawdust. Each 100 kg cultivation material can produce 100 kg of fresh mushrooms. In *Lentinula edodes* production, pomace can replace the same amount of woods, leading to excellent production efficiency and economic benefits. Optimal cultivation conditions include an inoculum amount of 10–15% cultivation materials, cultivation temperature of 25–27°C, and relative air humidity of 70%. Ze Xiong et al. used citrus peel for the cultivation of *P. ostreatus*, showing that the mushroom growth rate in citrus peel medium is 19 days, growth speed is 0.52 cm/day, and biological efficiency is 137%.

3.3 Other Biotransformation Products of Citrus Peel

3.3.1 Preparation of Citrus Peel Plastic Products

Citrus peel can be used in the plastics industry. The research team of Curtis from Cornell University has developed a catalyst to produce novel polymers through the combination between some major components in citrus peel and carbon dioxide. A new type of plastic made from citrus peel has properties similar to those of polystyrene, which is a widely used plastic.

As is well known, high amounts of carbon dioxide released into the atmosphere cause a rise in the temperature of the Earth. The researchers have demonstrated that this new technology provides a method not only for manufacturing plastics through renewable raw materials but also for "locking" carbon dioxide.

3.3.2 Production of Biogas

Citrus peel slag can be decomposed to produce biogas through anaerobic bacteria. According to reports, 90% of organic compounds can be biodegraded after 45 days fermentation to produce biogas with a methane content of up to 60%. Meanwhile, a ton of wet pomace can produce a net energy of up to 1.463–3.553 million kJ. The major production process of biogas is as follows:

Adjusting pH of pomace ⟶ Adding a certain amount of fresh dung ⟶ Water-sealed fermentation ⟶ Producing biogas.

A mixture of pomace and dung is placed in the fermentation pool to complete anaerobic fermentation for biogas production. After fermentation, the dehydrated residues can be used as high-quality fertilizers and feeds for fish and earthworms. Previous practice has proved that the application of biogas manure for rearing fish has multiple advantages over that of common fecal feeds. Recycling of the wastes produced can result in significant economic and social benefits.

3.3.3 Preparation of Citric Acid

After pectin or oil extraction, the pomace of citrus peel can be used as the raw material for preparing citric acid. In general, 100 kg of citrus peel waste can produce 215 g citric acid. Citric acid is a colorless, translucent crystal or white powder with a flavor of fruit acid. Citric acid is widely used in the food and beverage manufacturing industry as a food additive and in the oil processing industry as a souring agent, softener, or a preservative or antioxidant agent.

3.3.4 Preparation of Activated Carbon

After pectin extraction, the residues can be used to prepare activated carbon. The technological process is as follows:

Crushing dried citrus peel residue ⟶ Drying Carbonization ⟶ Smashing ⟶ Cooking ⟶ Washing ⟶ Centrifuging ⟶ Drying ⟶ Smashing ⟶ Sieving ⟶ Preparing activated carbon.

The carbonization temperature is 400°C. During cooking, 5% hydrochloric acid is added. The adsorption capacity of prepared products to methylene blue is 0.1494 g/g. In addition, Zhigang Xie et al. used citrus peel as the raw material and zinc oxide as an activator to prepare mesoporous activated carbon through a one-step method, and derived the optimal preparation conditions including a dipping ratio of citrus peel and an activator of 3:1, a carbonization temperature of 550°C, and a temperature-holding time of 1 h. Under the optimal preparation conditions, the pore volume of obtained activated carbon is up to 1.438 cm^3/g, porosity is 68.5%, the specific surface area is 1476 m^2/g, and the average pore size is 3.88 nm. Activated carbon is widely used in food, chemical, metallurgical, and environmental industries as a very important industrial product.

Although some functional components can be extracted from citrus peel for use in the processing of food and feeds, a large amount of citrus peel is wasted due to the lagging processing technology and processing equipment as well as unperfected regulations. However, with the increasing emphasis on environmental protection from governments and the growing awareness of environmental protection from people, all types of new processing technologies and new equipment will be applied to the comprehensive utilization of citrus peel, which will be benefited to reach a new scope and depth. Especially, with the development of biotechnology, various engineered bacteria with excellent performance will be used in the comprehensive utilization of citrus peel. In the long term, the application of biotechnology for the transformation of citrus peel should be a fundamental method for the comprehensive utilization of pomace of citrus peel.

CHAPTER 8

Production of Biodegradable Packages Using Citrus Peel

Contents

1. Technological Process 91
2. Key Operating Points 91
   2.1 Production of Throwing Trays 91
   2.2 Production of Nutritional Bowls 92

After extracting bioactive components, the pomace of citrus peel can be processed to prepare biodegradable packaging (molded) materials, such as egg trays, fruit trays, throwing trays, and other nutritional bowls, thus enabling the reutilization of citrus peel. These products do not result in environmental pollution through natural degradation, thus enabling full utilization without generation of citrus peel wastes.

1. TECHNOLOGICAL PROCESS

Citrus peel ⟶ Baking ⟶ Smashing ⟶ Pressing and molding

2. KEY OPERATING POINTS

2.1 Production of Throwing Trays

1. Pretreatment: Citrus fruits are cleaned and peeled, and the citrus peel is collected for future use.
2. Drying process: The pretreated citrus peel is placed in an oven for baking at 70–80°C for 8–10 h until complete dry.
3. Pulverization treatment: The completely dried citrus peel is placed in a miller, where it is crushed into powder at a crushing speed of 25,000 r/min.
4. Molding: The crushed citrus peel slag is mixed with a urea–formaldehyde resin at a ratio of 7:3 and placed on a hot extrusion mold to produce scattered-transplantation seedling plates.

Comprehensive Utilization of Citrus By-products
ISBN 978-0-12-809785-4
http://dx.doi.org/10.1016/B978-0-12-809785-4.00008-3

2.2 Production of Nutritional Bowls

1. Pretreatment: Citrus fruits are cleaned and peeled, and the citrus peel is collected for future use.
2. Drying process: The pretreated citrus peel is placed in an oven for baking at 70–80°C for 8–10 h until complete dry.
3. Pulverization treatment: The completely dried citrus peel is placed in a miller, where it is crushed into powder at the crushing speed of 25,000 r/min.
4. Molding: The crushed citrus peel slag is mixed with a urea–formaldehyde resin and starch at a ratio of 7:2:1 and placed on a large-scale high-temperature hot extrusion mold to produce nutritional bowls.

The scattered-transplantation seedling plates prepared by the described method can be naturally degraded in the soil and does not lead to environmental pollution. Similarly, the nutritional bowls can be used for crop cultivation.

CHAPTER 9

Processing Equipment for Citrus Peel By-products

Contents

1. Drying Machines and Equipment 93
 1.1 Pressure Spray Drying Equipment 93
  1.1.1 Principle of Pressure Spray Drying 94
  1.1.2 Types and Structures of a Pressure Sprayer 94
 1.2 Atmospheric Belt Dryer for Citrus Peel Feeds 96
 1.3 Box Freeze-Drying Equipment 97
  1.3.1 Working Principle 97
  1.3.2 Structure 98
  1.3.3 Operation Process 99
2. Recycling Equipment for Aromatic Substances in Citrus Juice 101
 2.1 Vacuum Crystal Aroma Recovery Equipment 101
 2.2 Multilevel Recovery Equipment for Aromatic Substances 102
 2.3 The Recovery Device for the Recovery and Vacuum Condensation of Aromatic Substances of Orange Juice 103
3. Typical Citrus Processing and Production Line 104
 3.1 Production Line of Citrus Pectin 104
 3.2 Production Line of Citrus Peel Feeds 105

1. DRYING MACHINES AND EQUIPMENT

At present, commonly used drying methods include spray drying, microwave drying, and freeze-drying. Spray drying and microwave drying are high-temperature and short-term drying methods. Freeze-drying is a low-temperature drying method. In China, the production of citrus juice powder and citrus crystals usually uses spray drying equipment.

1.1 Pressure Spray Drying Equipment

The spray drying technique is one of new modern drying technologies. Through mechanical action, the materials that need to be dried are dispersed as fine mist-like particles (improving the evaporation area and accelerating the drying process), and water in the materials is removed in a flash contacting with hot air. The raw materials can be dried as a powder

Comprehensive Utilization of Citrus By-products
ISBN 978-0-12-809785-4
http://dx.doi.org/10.1016/B978-0-12-809785-4.00009-5

directly through spray drying. Pressure spray drying is a simple and mature technology with high-production capacity. In the citrus processing industry, spray drying is often used in the production of juice powder, orange crystal, and other solid beverages.

1.1.1 Principle of Pressure Spray Drying

In pressure spray drying, a high-pressure pump is used to obtain a high-pressure liquid (7–20 MPa). The liquid enters the chamber of the nozzle from a tangential direction or from guide grooves of the nozzle core. At this time, part of the static pressure of the liquid can be converted to dynamic pressure energy to produce a strong rotational motion. According to the conservation law of rotation momentum, the rotation speed is inversely proportional to the radius of the vortex, which is closer to the axis and reveals the greater speed of the rotation, and the static pressure is smaller, thereby forming a central vortex air pressure equal to atmospheric pressure swirl, while the annular film of liquid is formed around the center of rotation. The static pressure of the liquid at the nozzle can be converted to kinetic energy for the forward movement of the film from the nozzle. Then, stretch film becomes thin and pulls into filaments. Because of the sudden loss of pressure, the liquid film can be split into droplets. A hollow conical spray moment can be formed by the droplet group, which is also called the hollow cone spray. In general, with higher spray pressure and smaller nozzle, smaller droplets are ejected; under contrast conditions, larger droplets are ejected.

1.1.2 Types and Structures of a Pressure Sprayer

A sprayer is an important component in spray drying equipment. The liquid material can be stably sprayed as droplets of uniform size and evenly distributed in the active portion of the drying chamber for keeping excellent contact with hot air.

At present the most-used pressure sprayers are Monarch (M) and Spraying (S) types.

1. M-type Sprayer: The structure of an M-type sprayer is shown in Fig. 9.1. It mainly consists of a pipe joint, nut, distribution plate, and nozzle. The nozzle is inserted with an artificial gem sprinkler head. An annular diversion ditch is opened at the side of the nozzle. The diversion ditch toward the nozzle has tangential channels composed of four horizontal grooves with various widths and depths on the basis of flux. Four horizontal grooves are perpendicular to the axis of the nozzle but do not intersect at the axis. On the distribution plate, four

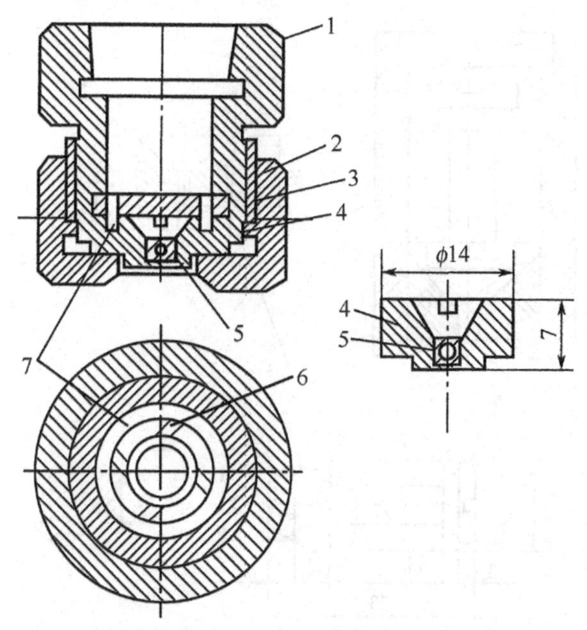

Figure 9.1 ***M-type pressure sprayer.*** (1) Pipe joint; (2) screw nut; (3) distribution plate; (4) nozzle; (5) sprinkler head; (6) tangential channel; (7) annular diversion ditch.

small holes are close to the relative annular diversion ditches. After the liquid is boosted through a high-pressure pump, the liquid enters annular diversion ditches through three small holes on the distribution plate. It then flows through four tangential channels on the side of the nozzle and is injected tangentially into the nozzle, resulting in a strong rotational motion. The formed annular film is sprayed from the nozzle. The liquid film or filaments under high pressure are split into tiny droplets and sprayed into the drying tower because of the sudden loss of the pressure for completing heat-mass exchange in hot air and dried instantaneously.

An M-type sprayer with a large flux is suitable for the drying equipment with a large production capacity. Ruby nozzles are utilized to obtain good wear resistance, a smooth orifice wall, consistent spraying status, and excellent product quality. Meanwhile, the large diameter of orifices of the ruby nozzles prevents blocking of the nozzles. Therefore, the operation is stable and the product quality can be easily controlled.

2. S-type Sprayer: The structure of an S-type sprayer is shown in Fig. 9.2. It consists of nuts, pipe joints, nozzles, and a spray core. On the spray

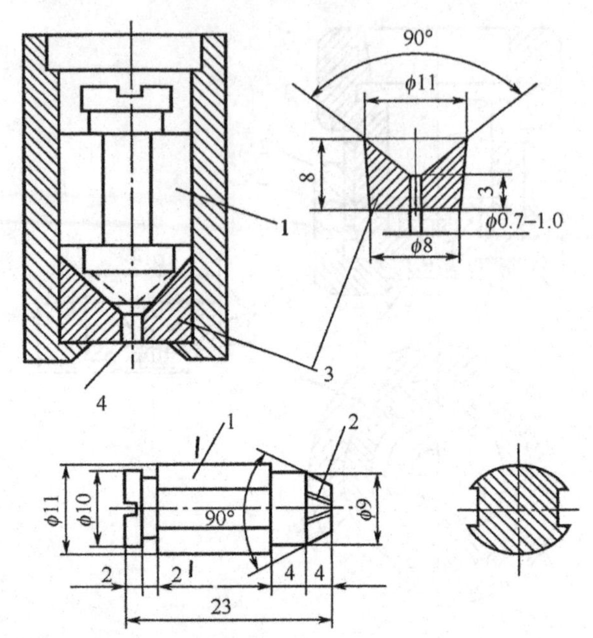

Figure 9.2 ***S-type pressure sprayer.*** (1) Spray core; (2) diversion groove; (3) nozzle; (4) nozzle hole.

core, there are two diversion grooves. The axis of the diversion groove has an angle with the axis of the nozzle core. These parts are made of stainless materials. The diameter of the nozzle is generally 0.5−1.4 mm.

1.2 Atmospheric Belt Dryer for Citrus Peel Feeds

Drying equipment involves a wide range of products. According to the pressure for classification, they include an atmospheric dryer and a vacuum dryer. According to the operation mode for classification, they include a continuous dryer and an intermittent dryer. An atmospheric dual-belt dryer is one such dryer.

An atmospheric belt dryer performs thermal transfer in a convection manner. The belt carrying citrus peel is moved in the drying room to complete the drying of citrus peel through contact with hot air. The atmospheric belt dryer is composed of a drying chamber, conveyor, fans, heaters, elevators, and an unloading machine. The belt commonly involves a canvas conveyor, steel belt, and steel mesh belt.

Fig. 9.3 presents a type of atmospheric belt dryer. The whole dryer is divided into two drying zones and a cooling zone. The first drying zone is

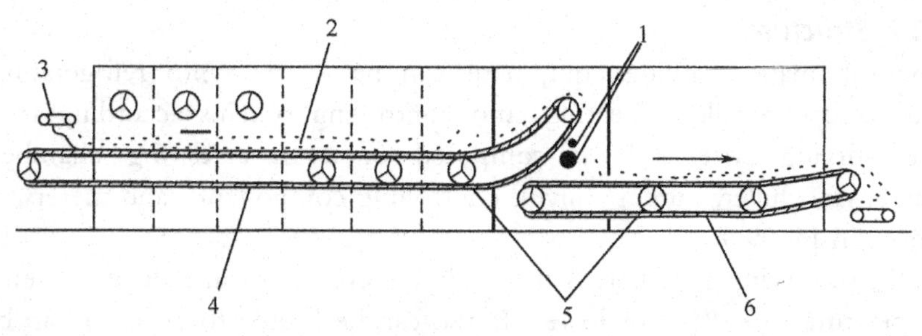

Figure 9.3 ***Schematic diagram of a dual-belt dryer.*** (1) Material-discharging roller and material-crushing roller; (2) material bed; (3) distributor; (4) first belt; (5) fan; (6) second belt.

divided into front and rear temperature zones. After citrus peel is dried at the first and second drying zones, it is transferred to the front part of the second belt from the end part of the first belt. During the transferring process, the materials are flipped and then passed through the cooling zone until the final discharge.

The advantages of such a dryer are as follows: (1) the citrus peel materials are transferred between belts and flipped to increase the evaporation area and to improve permeability and drying uniformity; and (2) the speeds of different conveyors can be controlled independently, the same as the speed of hot air flow, temperature, and humidity in multiple drying zones, which is helpful in optimizing the material drying process.

1.3 Box Freeze-Drying Equipment

1.3.1 Working Principle

The reduced pressure can result in reduction of the boiling point of water. When the pressure is reduced to 0.6 kPa and temperature is 0°C, or the pressure is 4.5 Pa and temperature is −50°C, the water in materials can directly sublimate into vapor from ice without passing through the liquid phase, which is the principle of freeze-drying. Freeze-drying is also known as sublimation drying. Vacuum freeze-drying is one kind of vacuum drying. The operating conditions of vacuum freeze-drying generally include a vacuum of 133.3−1.33 Pa and product temperature of −30°C to −50°C. Drying time in vacuum freeze-drying is longer than that in vacuum drying. Freeze-dried fruit powder is scaly or porous and beautiful. The solubility of freeze-dried fruit powder in water is excellent.

1.3.2 Structure

Box sublimation drying equipment can be divided into refrigerating, automatic controlling, heating, and controlling systems according to a freeze-drying system. It is composed of a freeze-drying chamber, condenser, chiller, pumps, valves, controlling components, and meters, as shown in Fig. 9.4.

1. Freeze-drying chamber: A freeze-drying chamber can reduce the temperature to −40°C or lower. It also can be heated to 50°C or can be evacuated. Usually, several layers of shelves in the chamber are set up. The chamber is connected with the condenser through a pipe fitted with a vacuum valve. The released vapor is transferred to the condenser through this pipe. In addition, several observation holes are present on the chamber, where the introduction plugs of electric wires for monitoring the temperatures of vacuuming freezing, the shelves, and the product are inserted.
2. Condenser: The condenser is a vacuum-sealed container with a large metal pipe surface. It can reduce temperature from −40°C to −80°C. After condensing, a large amount of vapor is discharged from the condenser to reduce the pressure in the chamber. It also has a defroster, exhaust valves, as well as hot air–blowing devices, which can be used for discharging ice and frost and drying the chamber.

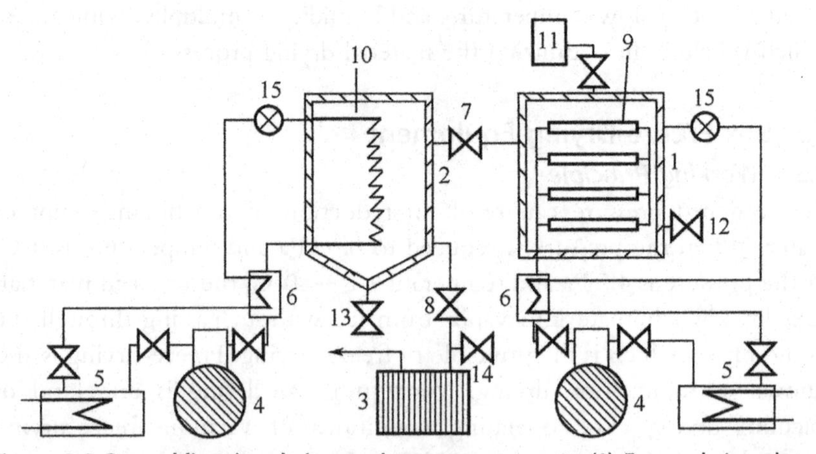

Figure 9.4 ***Box sublimation drying equipment components.*** (1) Freeze-drying chamber; (2) condenser; (3) vacuum pump; (4) refrigeration compressor; (5) water cooler; (6) heat exchanger; (7) condenser valve; (8) condenser vacuum valve; (9) plate temperature indicator; (10) condensing temperature indicator; (11) vacuum indicator; (12) freeze-drying bleeding valve; (13) gas condensate discharge outlet; (14) pump discharge valve; (15) expansion valve.

3. Vacuum pumps and vacuum gauges: The vacuum system is composed of a freeze-drying chamber, condenser, vacuum valves and pipes, vacuum pumps, and vacuum gauges and requires good sealing. Vacuum pumps usually include vane rotary pumps or oil-sealed mechanical pumps, which can be used together with a mechanical or oil boosting pump. A rotary mercury-squeezed vacuum gauge or resistance and a thermocouple vacuum gauge can be used.
4. Cooling system and heating system: A cooling system consists of freezers, a freeze-drying chamber, a condenser, and corresponding pipes. Freezers can be two independent sets, including the freeze-drying chamber and a freeze-drying condenser, or two sets combined into one. Freezing includes direct and indirect methods. The direct freezing system is used to transfer the refrigerant to the freeze-drying chamber or condenser. The freezer involves a single-stage, two-stage, or cascade compressing freezer according to the demands of low temperature. A refrigerating compressor can be an ammonia or Freon compressor.

   A heating system is used to heat the clapboards in the freeze-drying chamber for promoting the sublimation of products, including direct and indirect methods. The direct heating method uses direct electric heating in the chamber. The indirect heating method utilizes electricity or other heating sources to heat transfer the medium for passing into the clapboards.
5. Control system: A complex freeze-drying equipment control system includes various switches and safety devices, which make the system highly automated.

1.3.3 Operation Process

1. Prefreezing: Prefreezing is a process used to decrease the temperature and speed. Because materials contain a large amount of water, dissolved gas in the water can quickly form bubbles and be released by the outside pressure reduced, and leaves the materials a "boiling" status. Water evaporates into steam and absorbs its own heat to form ice. The ice is regasified to produce foaming and air-agitating products with more holes inside. In general, prefreezing can reduce the temperature to −30°C. Different materials have different eutectic points. A eutectic point is the temperature required to freeze the material to form a solid. It requires prefreezing temperature below the eutectic point of approximately 5°C. If the temperature cannot meet the

requirements, freezing will be incomplete. Prefreezing time should be controlled at 2 h on the basis of the starting point below the eutectic point because each shelf needs to be given sufficient time as they have different temperatures. In addition, prefreezing speed should be controlled at 1–4°C/min. Too high or too low prefreezing speed is detrimental to the product. The prefreezing speeds of different products are determined by practical trials.

2. Sublimation process: After prefreezing and evacuating, the temperature is almost constant at the second stage. The process is a constant speed process with the discharge of frozen water. Direct vaporization of ice requires heat absorption. At this time, heating can be started to keep the temperature close to but below the eutectic point. If heating is not provided, the temperature of the materials will drop, thus leading to a reduction in drying speed, prolonged drying duration, and unqualified water content in products. In addition, too much or excessive heating can result in the increase in temperature of the material itself. If the temperature of the material itself is higher, it will exceed the eutectic point and result in partial melting and reduced volume, as well as blistering of the material. For example, 1 g of ice can produce 9500 L vapor at a pressure of 13.3 Pa. It is impossible for a common mechanical pump to remove the vapor with a bulky volume. However, the steam jet pump requires high-pressure steam and multilevel cascades for pumps, which is uneconomical for small and medium enterprises. Therefore condensers can be applied to cool the vapor on the surface, thus forming frost. If the temperature is held at −40°C, vapor pressure in the condenser can be reduced to a certain level, while the vapor pressure in the oven is increased. The pressure difference can accelerate the vapor to continuously transfer to the condenser.
3. Evaporation of remaining water: The heating rate can be accelerated to evaporate the unfrozen water, once the frozen water is fully evaporated and the products have been shaped. At this time, the drying rate is decreased and the water is continuously evaporated. Although the temperature can be gradually increased; it is generally not more than 40°C. The optimal temperature is 30–35°C, with a temperature-holding duration of 2–3 h. After drying is completed, the final products are collected. The condenser can be heated under atmospheric pressure to discharge the water from the melted ice and frost.

2. RECYCLING EQUIPMENT FOR AROMATIC SUBSTANCES IN CITRUS JUICE

Fresh citrus juice contains various aromatic substances. If the original citrus juice is subjected to direct evaporation, aromatic substances will escape with the secondary steam, which will result in the loss of aromatic flavor in the concentrated citrus juice. Therefore, during condensation of citrus juice, it is necessary to first extract aromatic substances, prepare the fragrant liquid, and then convert it to the concentrated citrus juice for keeping the original citrus flavor.

2.1 Vacuum Crystal Aroma Recovery Equipment

Fig. 9.5 shows the schematic diagram of vacuum crystal aroma recovery equipment. When juice enters the vacuum freezer, it is evaporated and frozen under an absolute pressure of 267 Pa. Partial water is converted into ice crystals. The ice crystals coming from the freezer are separated from the frozen suspension in the separator. The concentrated solution enters the upper part of the absorber and is discharged from the bottom of the absorber. The aromatic substances—containing vapor coming from the

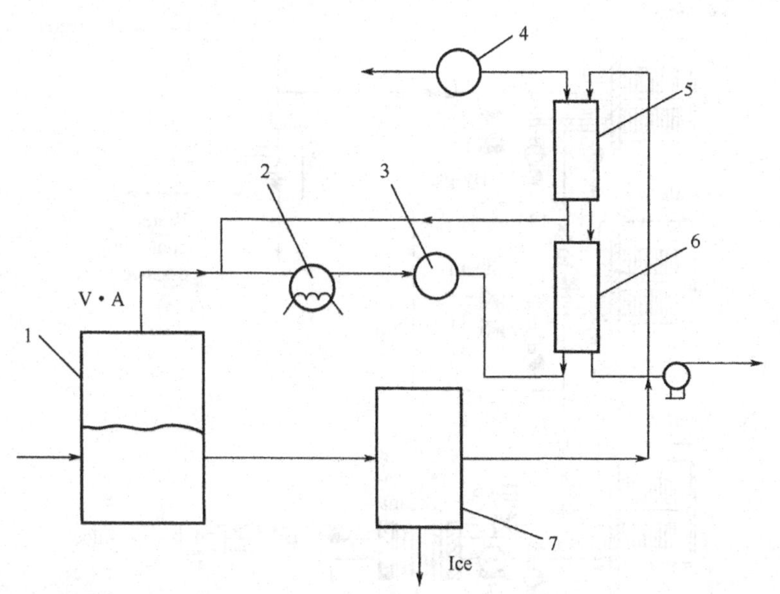

Figure 9.5 ***Schematic diagram of vacuum crystal aroma recovery equipment.*** (1) Vacuum crystallizer; (2) condenser; (3) dry-style vacuum pump; (4) wet-style vacuum pump; (5) absorber II; (6) absorber I; (7) ice and crystal separator; (V) vapor; (A) aromatic substances.

vacuum crystallizer passes through the condenser, where water is removed, and then enters the absorber from the bottom, and the inert gas is extracted from the upper part of the absorber. The concentrated solution flows reversely with the inert gas containing aromatic substances in the absorber. If the temperature of the condenser is not too low, to further reduce the loss of aromatic substances, the inert gas that has left the first absorber can be returned to the condenser for recycling.

2.2 Multilevel Recovery Equipment for Aromatic Substances

In addition to the abovementioned single-stage recovery method, the multilevel recovery method is applied for the extraction of aromatic components from juice. At present the recovery equipment for aromatic substances in juice processing factories in China are of two- or three-stage types. Fig. 9.6 shows the technological process of the three-stage aromatic component recovery equipment from Swedish Alfa-Laval Company. The working process is as follows: the juice passes through the heater for pre-heating treatment and then enters the evaporator through heating by steam

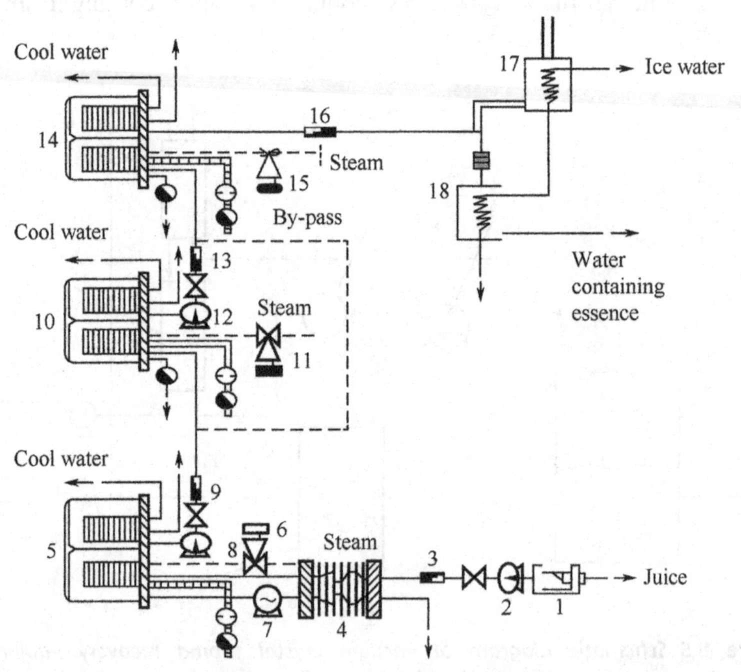

Figure 9.6 ***Schematic diagram of a three-stage aromatic component recovery equipment.*** (1) Balance tank; (2, 7, 8, 12) centrifugation pump; (3, 9, 13, 16) flowmeter; (4) plate heating exchanger; (5) PAR-22 aroma extractor; (6, 11, 15) steam control valve; (10, 14) PAR-01 aroma extractor; (17, 18) spiral condenser.

to produce the secondary steam that contains volatile aromatic components. The secondary steam is brought to the condenser through baffles and then cooled into a fragrant liquid by the cooling water in the heat exchanger. The fragrant liquid in the two- and three-stage aromatic collectors is concentrated through evaporation and condensation to improve the concentration of the fragrant liquid. The extracted juice and aromatic condensate are discharged from the lower part of the device, and the juice is brought to the evaporator for condensation. The specifications of the commonly used two- or three-stage aroma recovery equipment are as follows:

PAR-22 (Primary aromatic collector): Amount of water evaporated is 1500 kg/h. The primary amount of aromatic components collected is 10–25% of the total amount of fragrant juice.

PAR-01 (two- or three-stage aromatic collector): Amount of water evaporated is 140 kg/h. The collection of aromatic components by the two-stage collector is 10–25% of balsam water, and the highest steam temperature is 120°C.

If the amount of juice (or balsam) evaporated each time is 10%, the essence obtained by the two-stage collector is usually called 100-fold juice essence. Similarly, the essence refined by the three-stage collector is usually called 150-fold juice essence.

2.3 The Recovery Device for the Recovery and Vacuum Condensation of Aromatic Substances of Orange Juice

During the condensation of citrus juice, if a thin-film evaporator is used, volatile aromatic components of citrus juice escape easily. To improve the flavor quality of concentrated orange juice, aromatic components of citrus juice should be recovered. Fig. 9.7 shows the flowchart of the recovery of aromatic components of juice and a vacuum condenser. Original fruit juice is transferred to a filter through the juice pump for filtration. The filtered juice is subjected to preheating in a preheater and heating in a heater. Then, the juice enters a flash steamer where vapor, aromatic substances, and defragrant juice are separated. Vapor and aromatic components enter the aroma water separator for vapor condensation. The aromatic vapor in the condenser and essence in the refrigerator are cooled as liquid to obtain the recovered essence. After separation in the flash steamer, the defragrant juice in the evaporator is concentrated. The generated secondary steam is condensed, which then becomes a condensate after passing through a steam separator. The concentrated juice is discharged through the pump. The degree of vacuum in the whole system is maintained by a steam-injecting pump.

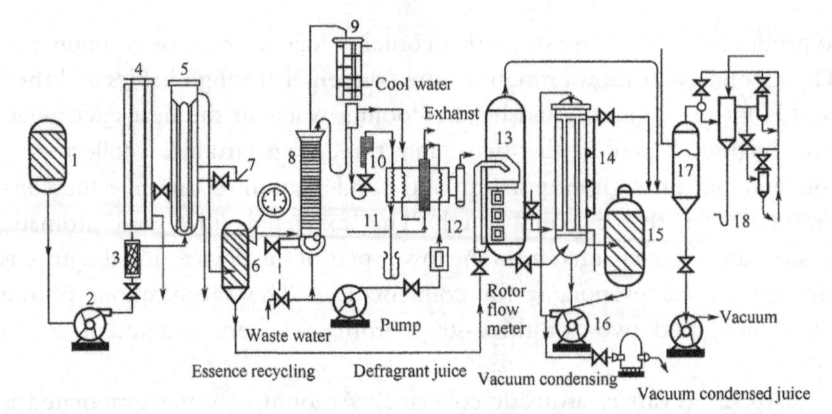

Figure 9.7 ***Schematic diagram of the recovery of aromatic components during juice and vacuum concentration.*** (1) Juice container; (2) juice pump; (3) filter; (4) preheater; (5) heater; (6,13) flash steamer (instant evaporator); (7) heating steam inlet; (8) aromatic water separator; (9) condenser; (10) essence cooler; (11) essence depth cooling system; (12) ammonia freezer; (14) evaporator; (15) steam separator; (16) circulating pump; (17) vacuum holder; (18) steam-injecting pump.

3. TYPICAL CITRUS PROCESSING AND PRODUCTION LINE

In this chapter, two typical citrus processing and production lines have been described for readers' reference (Figs. 9.8 and 9.9).

3.1 Production Line of Citrus Pectin

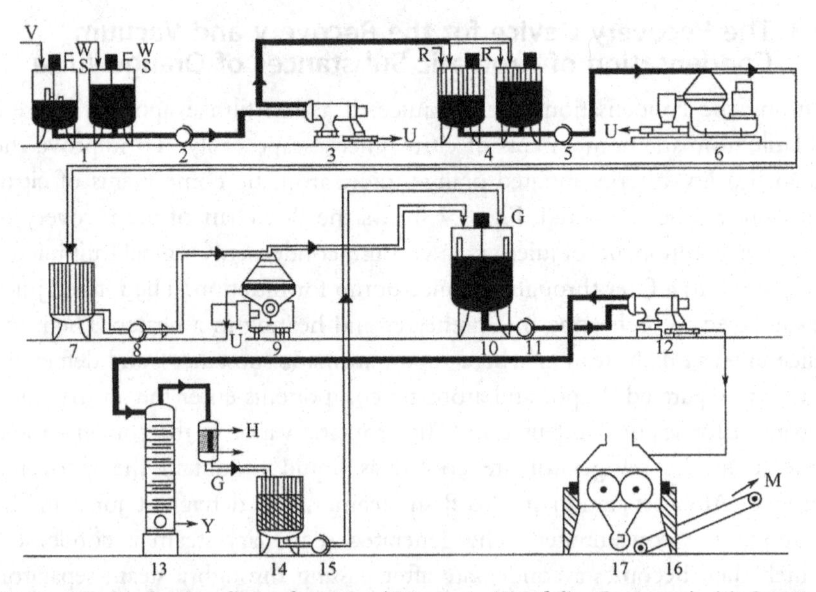

Figure 9.8 ***Production line of pectin (Germany, Westfalia Company).*** (1) Reactor; (2) pump; (3) separator; (4) buffer tank; (5) pump; (6) preclarifer; (7) buffer tank; (8) pump; (9) high efficient clarifier; (10) presedimentation tank; (12) separator; (13) distillator; (14) buffer tank; (15) pump; (16) drum dryer; (17) miller; (G) ethanol; (H) cooling water; (M) pectin; (R) clarifying agent; (S) acid; (U) solid; (V) citrus peel; (W) water; (Y) waste.

3.2 Production Line of Citrus Peel Feeds

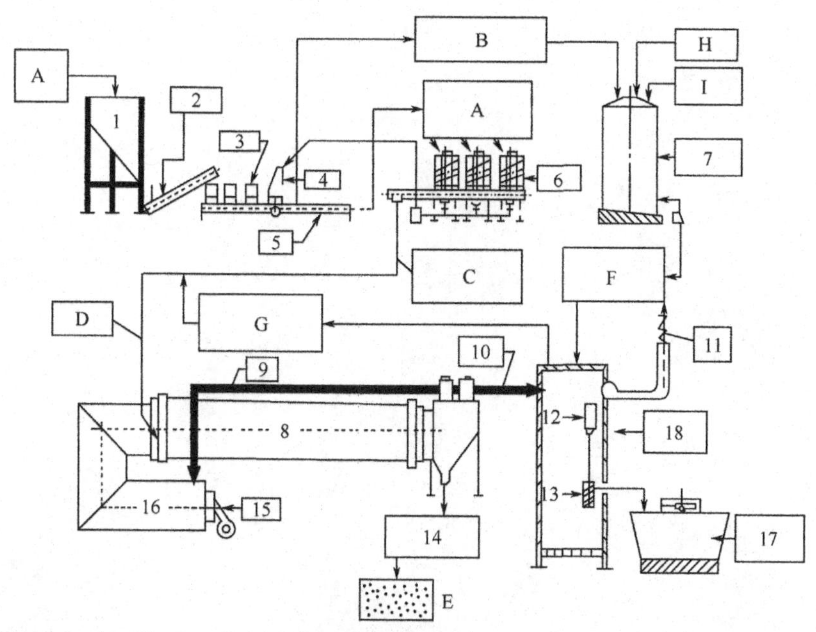

Figure 9.9 ***Production line of citrus peel feeds and raw material balance.*** (1) Citrus peel container; (2) lime; (3) crusher; (4) separating sieve; (5) spiral reactor; (6) pressor; (7) condenser feeding container; (8) direct heating dryer; (9) circulating steam; (10) exhaust passage of the dryer; (11) waste gas; (12) Baume meter; (13) heating well; (14) juice cooler; (15) fuel; (16) burner; (17) cooler; (18) condenser; (A) pomace, water content of 82%, 65,409 kg/h; (B) condensate, 9°Brix, 34,405 kg/h; (C) pressed cake, water content of 72%, 31,004 kg/h; (D) raw materials in a dryer feeding inlet, water content of 70.3%, 40,629 kg/h; (E) granular feed products, water content of 10%, 13,389 kg/h; (F) condenser feeding, 6.16a°Brix, 55,024 kg/h; (G) concentrated molasses, 35°Brix, 9625l kg/h; (H) waste liquid from essential oil factory, 2°Brix, 13,620 kg/h; (I) wastewater from factory, 7000 kg/h.

9.2 Production Line of Citrus Peel Feeds

Figure 9.5 Production line of citrus peel feeds and raw matter of balance. (1) [illegible] (2) [illegible] (3) [illegible] (4) separating [illegible] (5) [illegible] (6) [illegible] (7) [illegible] feeding conveyer, (8) direct heating dryer, (9) [illegible] steam, (10) exhaust [illegible] of the dryer, (11) waste gas, (12) [illegible] meter, (13) [illegible] well, (14) [illegible] cooler, (15) [illegible] (16) [illegible] (17) [illegible] (18) condenser. (A) [illegible] water content of 82%, 65,[illegible] kg/h; (B) condensate [illegible] 34,403 kg/h; (C) pressed cake water content of 79%, [illegible] kg/h; (D) [illegible] in a [illegible] water content [illegible] kg/h; (E) [illegible] feed products, water content of 10%, 12,[illegible] kg/h; (F) condensate [illegible] (G) concentrated molasses, [illegible] Brix 95, [illegible] kg/h; (H) waste liquid from essential oil factory, [illegible] Brix, 13,620 kg/h; (I) wastewater from factory [illegible] kg/h.

Appendix

1. STANDARD CATALOG OF CITRUS AND PRODUCTS FROM ABROAD

No.	Standard no.	Titles
1	ISO 8901:2003	Oil of bitter orange petitgrain, cultivated (*Citrus aurantium* L.)
2	ISO 3809:2004	Oil of lime (cold pressed), Mexican type [*Citrus aurantifolia* (Christm.) Swingle], obtained by mechanical means
3	ISO 3053:2004	Oil from grapefruit (*Citrus* × *paradisi* Macfad.), obtained by expression
4	ISO 855:2003	Oil of lemon [*Citrus limon* (L.) Burm. f.], obtained by expression
5	ISO 8898:2003	Oil of mandarin petitgrain (*Citrus reticulata* Blanco)
6	ISO 8899:2003	Oil of lemon petitgrain [*Citrus limon* (L.) Burm. f.]
7	ISO 8900:2005	Oil of bergamot petitgrain [*Citrus bergamia* (Risso et Poit.)]
8	ISO 3519:2005	Oil of lime distilled, Mexican type [*C. aurantifolia* (Christm.) Swingle]
9	NY/T2010-2011	Determination of total flavonoids in citrus fruits and products
10	NY/T2011-2011	Determination of limonin in citrus fruits and products
11	NY/T2013-2011	Determination of essential oil in citrus fruits and products
12	NY/T2014-2011	Determination of hesperidin and naringin in citrus fruits and products

REFERENCES

[1] Yang S. Introduction of citrus processing. Beijing: China Agriculture Press; 2004.
[2] Herong L, Yang S. Practical fruit processing technology. Changsha: Hunan Science and Technology Press; 1997.
[3] Yang S. Separation, structural identification, chemical modification and biological activity of flavonoids from citrus (Ph.D. thesis). Changsha: Central South University of Forestry and Technology; 2007.
[4] Yang S. Current status and development trend of fruit processing industry in China. Journal of Beijing Technology and Business University (Natural Science Edition) 2012;30(3):1–12.
[5] Yang S. Current status and functional properties of fruit and vegetable processing industry in China. Agricultural Engineering Technology (Agro-Product Processing Industry) 2010;(6):6–8.
[6] Yang S. Current status and strategic development thinking of fruit and vegetable processing industry in China. Journal of Chinese Institute of Food Science and Technology 2010;10(1):1–9.
[7] Yang S. Current status, development trend and strategy of citrus industry in China. Journal of Chinese Institute of Food Science and Technology 2008;8(1):1–8.
[8] Yang S, Li G, Qiuan W. The semi-synthesis of five kinds of bioactive flavonoids from hesperidin. Organic Chemistry 2008;(6):1024–8.
[9] Yang S, Li G. Research progress of modern biotechnology in citrus industry. Food and Machinery 2007;23(5):142–5.
[10] Yang S, Li G, Li Z. Antioxidant activity in vitro of methoxyflavone in citrus peel. Food Science 2007;28(8):100–3.
[11] Shan Y. Current status and research and development system of citrus industry in Spain. Food and Machinery 2004;20(1):48–9.
[12] Shan Y, Li W, He J. Current status and development of citrus industry in Brazil. Journal of Hunan Agricultural University (Social Science Edition) 2004;5(6):1–5.
[13] Yang S, He J, Fu F. Current status, strategies and prospects of citrus industry in Hunan province. Hunan Agricultural Sciences 2003;(5):58–61.
[14] Yang S, He J, Li G. Current status, development prospects and strategies of agricultural commodity industry in Hunan. Research of Agricultural Modernization 2003;24(4):303–7.
[15] Yang S, He J. Citrus processing technology in Hunan. Hunan Agricultural Sciences 1999;(3):36–7.
[16] Yang S, Herong L. Comprehensive utilization of Pomelo. Hunan Agricultural Sciences 1997;(2):40–1.
[17] Yang S, Li G, He J. The extraction method of low methoxy pectin from citrus using pectinase: China, CN200710035847.3 [P].2008-04-02.
[18] Yang S, Li G, Fu F. The preparation methods of dietary fiber nutrition powder using citrus sac pomace: China. June 27, 2012. CN201110389768.9 [P].
[19] Li C. Enzymatic preparation and characteristics of low methoxy pectin (Master thesis). Changsha: Central South University; 2010.
[20] Chunfu T. Extraction, purification and determination of phenolic acids from citrus peel (Master thesis). Changsha: Hunan Agricultural University; 2010.
[21] Sun D. Extraction, separation and purification of flavonoids from citrus peel (Master thesis). Changsha: Hunan Agricultural University; 2007.
[22] Ding X. Antioxidant activity and healthcare functions of the extract from citrus peel (Ph.D. thesis). Chongqing: Southwest Agricultural University; 2004.

[23] Su D, Li G, He J. Application of near infrared spectroscopy in the quality detection of a large amount of fruits in China. Science and Technology of Food Industry 2012;33(6):460–4.
[24] Su D, Li G, He J. Component analysis of amino acids and fatty acids in Satsuma mandarin peel. Science and Technology of Food Industry 2012;33(2):100–2.
[25] He D, Yang S, Wu Y. Simultaneous determination of flavanones, hydroxycinnamic acids and alkaloids in citrus fruits by HPLC-DAD-ESI/MS. Food Chemistry 2011;127(2):880–5.
[26] Su D, Li G, Liu W. Optimal process of bleaching and condensation of pectin from citrus fruits. Journal of Chinese Institute of Food Science and Technology 2011;11(4):108–14.
[27] Su D, Li G, Liang C. Optimal production process and quality analysis of pectin from orange peel. Food Science 2011;32(18):95–101.
[28] Xuebo S, Zhang J, Yang S. GC-MS analysis of volatile essential oil from citrus peel. Food Science 2010;31(2):175–8.
[29] Li C, Yang S, Li G. Optimization of enzymatic preparation process of low methoxy pectin. Foods and Machinery 2010;(2):110–4.
[30] Chunfu T, Yang S, Li G. Ultrasonic-assisted extraction process of phenolic acids from citrus peel. Science and Technology of Food Industry 2010;(8):223–4. 229.
[31] Fu F, Li Z, Yang S. GC-MS analysis of three kinds of citrus peel essential oil. Food and Machinery 2010;(3):30–4.
[32] Wu Y, Yang S, Deming X. Development proposal of citrus industry in Huahua. Hunan Agricultural Sciences 2009;(2):102–4. 107.
[33] Fu F, Li Z, Yang S. Research progress of comprehensive utilization of citrus peel. Food and Machinery 2009;(5):178–84.
[34] Su D, Yang S, Li G. Optimization of ultrasonic extraction of flavonoids from citrus peel by response surface methodology. Chinese Journal of Spectrosocopy Laboratory 2009;(6):1391–7.
[35] Sun D, Yang S, Li G. Optimization of the extraction process of total flavonoids from citrus peel by response surface methodology. Journal of Chinese Institute of Food Science and Technology 2009;(3):70–7.
[36] Su D, Yang S, Li G. Optimization of enzymatic-assisted extraction of total flavonoids from citrus peel. Transactions of the Chinese Society of Agricultural Engineering 2008;24(4):240–5.
[37] Su D, Yang S, Li G. Simultaneous determination of hesperidin and naringin in citrus peel by RP-HPLC. Science and Technology of Food Industry 2008;(6):288–90.
[38] Su D, Yang S, Li G. Optimization of extraction process of flavonoids from citrus peel and its mathematical model. Food Science 2008;(5):167–72.
[39] Liu W, Yang S, Li G. Determination of naringin in citrus fruits. Food and Machinery 2008;(6):102–6.
[40] Su D, Yang S, Li G. Determination of total flavonoids in citrus peel by colorimetry. China Brewing 2008;(3):69–74.
[41] Su D, Yang S, Li G. UV spectrophotometric determination of total flavonoids in citrus peel. Food Research and Development 2007;28(8):124–8.
[42] Su D, Yang S, Li G. Microwave extraction of total flavonoids from citrus peel. Food and Machinery 2007;23(3):73–5. 115.
[43] Su D, Yang S, Li G. Research progress in the extraction process of physiologically active components from citrus peel. Natural Product Research and Development 2007;(19):187–92.
[44] Su D, Yang S, Li G. Research progress in extraction process of functional compounds in orange peel. Modern Food Science and Technology 2007;23(3):90–4.

[45] Su D, Yang S, Li G. Research progress in extraction process of physiologically active components in citrus peel. The Beverage Industry 2006;9(11):6–10.
[46] Su D, Yang S. The overview of functional components in citrus peel. Modern Food Science and Technology 2006;22(2):260–2. 256.
[47] Qiuan W, Zhou B, Yang S. Extraction methods and antioxidant activity of flavonoids and natural extraction method. Chemical Production and Technology 2004;11(5):29–32.
[48] Tang J, Qiuan W, Yang S. Ultrasonic extraction of hesperidin from citrus peel. Fine Chemicals 2004;21(3):171–3.
[49] Qiuan W, Zhou B, Yang S. Extraction of carotenoids, hesperidin, pectin and limonoids from citrus processing wastes. Journal of Chemical Industry of Forest Products 2004;38(3):33–7.
[50] Qiuan W, Yang S. Extraction and modulation of essential oil from citrus peel. Science and Technology of Food Industry 2002;23(6):41–3.
[51] Qiuan W, Yang S. Comprehensive utilization of citrus peel resources. Recycling Research 1999;(5):28–30.
[52] USDA Foreign Agricultural Service. Citrus: World Markets and Trade [R/OL]. July 26, 2012. http://www.fas.usda.gov/report.asp.
[53] Zhang F. Statistics and analysis of processed fruit trade in China in 2010. China Fruit News 2011;28(3):1–9.
[54] Zang Y, Shao J. Research progress in citrus peel biotransformation. Journal of Chengde Petroleum College 2011;13(1):40–3.
[55] Cajustea JF, González-Candelasa L, Veyrata A, et al. Epicuticular wax content and morphology as related to ethylene and storage performance of 'Navelate' orange fruits. Postharvest Biology and Technology 2010;55(1):29–35.
[56] Wei R. The road from the big country of fruit and vegetable industry to the strong country of fruit and vegetable industry. Productivity Research 2006;3:51–2.
[57] Ge Y, Chen Y, Zhang Z, et al. Opinion in the development of fruit and vegetable processing industry in China. Food Science 2005;26(7):270–4.
[58] Hu X, Liao X, Chen F, et al. Current status and development trend of fruit and vegetable processing industry in China. Food and Machinery 2005;21(3):4–9.
[59] Li S. Current status of fruit production in the world and strategies for improving the competitiveness of fruit market in China. Journal of China Agricultural University 2003;8(1):7–13.
[60] Li S. Classification of functional components in citrus peel and green co-production process. Zhejiang Citrus 2003;20(3):25–8.
[61] Liu X, Deng X, Wang X, et al. Brazilian citrus inspection report. South China Fruits 2003;32(5):20–6.
[62] Deng X. Import and export status and development trend of citrus fruits and related products. World Agriculture 2001;10:23–5.
[63] Li W, Shao W, Gong M. Preparation of health composite orange peel sauce. China Condiment 2001;5:20–1.
[64] Houjiu W. Current status and development of citrus industry in China. Southern China Fruits 2001;30(4):19–20.
[65] Fan Z, Zhang X. Comprehensive utilization of citrus peel. Adaptation Technology Market 2000;12:39–40.
[66] Tang C, Zhiying P. Studies on functional components in citrus fruits. Sichuan Food and Fermentation 2000;4:1–7.
[67] Zhang H, Li J. Continuous extraction process of pigments and pectin from citrus peel. Food Science 2000;21(11):37–40.
[68] Houjiu W, Bilin J. Studies on citrus peel fermented feeds. China Feed 1997;17:37–9.
[69] Zhou Z. Citrus fruit nutrition. Beijing: Science Press; 2012.

[70] Chen X. Manual for exported citrus safety and quality control. Beijing: China Agriculture Press; 2009.
[71] Houjiu W. Processing and comprehensive utilization technology of citrus fruits. Chongqing: Chongqing Publishing House; 2007.
[72] Li S, Zhang Q. Food processing machinery and equipment manual. Beijing: Science Press; 2006.
[73] Agro-industry Leading Group Office from Ministry of Agriculture. Screening studies on major key agro-processing technology. Beijing: China Agriculture Press; 2006.
[74] Zhu B. Practical food processing technology. Beijing: Chemical Industry Press; 2005.
[75] Ye X. Citrus processing and comprehensive utilization. Beijing: China Light Industry Press; 2005.
[76] Development and Planning Department From Chinese Ministry of Agriculture. Regional layout and plan of competitive agricultural products. Beijing: China Agriculture Press; 2005.
[77] Cui J. Food processing machinery and equipment. Beijing: China Light Industry Press; 2004.
[78] Chen B. Food processing machinery and equipment. Beijing: Mechanical Industry Press; 2003.
[79] Zhu B. Feed production technology and equipment manual. Beijing: Chemical Industry Press; 2003.
[80] Standardization Management Committee in China. GB/T 5009-2003 China standard book. Beijing: China Standard Press; 2003.
[81] Agricultural Department of Hunan Province. Review and prospect of citrus industry in Hunan. Changsha: Hunan Science and Technology Press; 2001.
[82] Liu C, Zhou Y. Handbook of food additives. Beijing: Beijing University Press; 2000.
[83] Braddock RJ. Handbook of citrus byproducts and processing technology. A Wiley-Interscience Publication. John Wiley & Sons, Inc.; 1999.
[84] Liu Z. Food additives manual. Beijing: China Light Industry Press; 1996.
[85] Wu W. Agricultural processing engineering design. Beijing: China Light Industry Press; 1994.
[86] Li S. Contemporary citrus. Chengdu: Sichuan Science and Technology Press; 1990.
[87] Wuxi Institute of Light Industry. Shanghai Light Industry Design Institute From Ministry of Light Industry. Fundamentals of Food Factory Design. Beijing: China Light Industry Press; 1990.
[88] Shanwen H. Citrus manual. Changsha: Hunan Science and Technology Press; 1988.
[89] Wuxi University of Light Industry and Tianjin Institute of Light Industry. Food factory machinery and equipments. Beijing: China Light Industry Press; 1981.

INDEX

'*Note*: Page numbers followed by "f" indicate figures and "t" indicate tables.'

A

Acrylketones, 12f, 13
Atmospheric belt dryer, 96–97, 97f

B

Biodegradable packages
  nutritional bowls production, 92
  technological process, 91
  throwing trays production, 91
Biotransformation
  activated carbon preparation, 89–90
  biogas production, 89
  citric acid preparation, 89
  citrus peel plastic products preparation, 88–89
  ethanol, 87–88
  food production
    citrus vinegar, 85–86
    lactic acid drinks, 86
    natural turbidity agent, 86
  high-protein feeds, 86–87
    technological process, 87f
  mushroom cultivation, 88
Box freeze-drying equipment
  operation process, 99–100
  structure, 98–99, 98f
  working principle, 97

C

Chromone, 4, 5f
Cold milling process
  centrifugation, 37
  cleaning, 36
  condensation and terpene removal, 38
  dermabrasive oil mill, 36–37
  filtration, 37
  packaging and preservation, 38
  standing, 38
Cold-pressed oil
  cleaning, 39
  crushing, 39
  material selection, 38
  quality standards, 39, 40t
  soaking lime, 38–39
Coumarin, 12

D

Determination methods
  aromatic components, 24–29
    data processing, 25–29, 27t–28t
    GC–MS conditions, 25, 26f
    headspace solid phase microextraction conditions, 25
    volatile oils extraction, 24–25
  chemical analysis
    calculation, 18
    notes, 18
    pectin extraction. *See* Pectin extraction
    principle, 16
    reagents, 16
    sample handling, 16–17
  HPLC-nuclear magnetic resonance (NMR), 29–30
  liquid chromatography–mass spectrometry (LC–MS), 29
  multi-methoxyflavones, high-performance liquid chromatography, 20–24
  spectral analysis, 18–20
Diode array detector (DAD), 20–21
Drying, citrus peel
  operating points
    drying, 77
    manual sorting, 77
    mechanical screening, 77
    packaging, 77
    pesticides removal, 76
    raw material, 76
    rinsing, 77

Drying, citrus peel (*Continued*)
segmentation, 77
storage, 77
wind and magnetic sorting, 77
quality requirements, 77–78
technological process, 76
Dual-belt dryer, 97f

E

Electrospray ionization-MS (ESI-MS), 25
Equipment processing
aromatic substances, recycling equipment for, 101–103, 101f–102f, 104f
box freeze-drying equipment
operation process, 99–100
structure, 98–99, 98f
working principle, 97
drying machines, 93–100
M-type sprayer, 94–95, 95f
pressure spray drying equipment, 93–96
principle, 94
types and structures, 94–96
production
citrus pectin, 104–105, 104f
citrus peel feeds, 105, 105f
S-type sprayer, 95–96, 96f
Extraction processes
edible cellulose powder, 54–55
bleaching/filtration and drying, 55
crushing and sieving, 55
grinding and soaking, 55
product quality indexes, 55
technological process, 55
water immersion, 55
essential oils
bergamot oil, 35
bitter orange oil, 34
cold milling method, 35
cold milling process. *See* Cold milling process
cold-pressed method, 35–36
cold-pressed oil. *See* Cold-pressed oil
d-limonene, 39–41
grapefruit oil, 35
lemon oil, 34
lime oil, 34
mandarin oil, 34
orange oil, 33
tangerine oil, 34
water distillation method, 36
flavonoids, 49–51
hesperidin. *See* Hesperidin
naringin, 51
limonoid, 51–53
limonin extraction. *See* Limonin extraction
separation and purification, 53
pectin, 41–48
high methoxyl pectin. *See* High methoxyl pectin
low methoxyl pectin. *See* Low methoxyl pectin
quality standards, 48, 48t–49t
seed oil
cleaning and drying, 56
grinding, 56
oil extraction, 56–57
quality standards, 58, 58t
refining, 57–58
solvent extraction, 57
synephrine, 53–54
percolation conditions, 54
product quality indexes, 54
sample treatment, 54
separation and purification, 54
technological process, 53
water overnight soaking, 54

F

Feed additives, 80–83, 81t
citrus peel powder
animal feeds, 82
nutrients, 82
pros and cons, 82
technological process, 82
citrus pulp silage
animal feeding, 83
nutrients, 83
pros and cons, 83
technological process, 82
dried citrus peel

nutrient scores, evaluation of, 81
pros and cons, 81
technological process, 81
Flavanone, 5
Flavonoids
compounds
diosmetin, 68
diosmin, 68
7-*O*-farnesyl hesperetin and 7,3-*O,O*-bifarnesyl hesperetin, 70
hesperetin, 66–67, 67f
luteolin, 69
persicogenin, 67–68
7-*O*-prenyl Hesperetin, 69
compounds synthesis, 71–72, 71f
7-*O*-prenyl Hesperetin/7-*O*-farnesyl hesperetin and 7,3-*O,O*-bifarnesyl hesperetin, 72–73
separation and purification, technological process for, 60–62, 61f, 63f–64f
ethanol extraction treatment, 61
extraction, 60–61
qualitative test methods, 60
separation, 62
structural identification
HPLC-MS, 62
identification, 63
NMR technique, 63
synthesis routes, 66, 67f
Functional components
carotenoids, 8–10, 10f
cellulose, 10–11, 11f
essential oil, 1–2
flavonoids, 4–8, 6f, 6t
limonoids, 8, 9f
monoterpenes, 2–3, 2f
other components, 12–13
pectin, 3–4, 3f
synephrine, 11–12, 12f

H

Hesperidin, 7
cooling and precipitation, 50
lime treatment, 49–50
neutralization and keeping warm, 50
quality standards, 51
raw materials, 49
refining, 50–51
separation, 50
squeezing/precipitation and filtration, 50
High methoxyl pectin
acidic extraction, 42–43
condensation, 43
decoloration, 43
drying, 44
filtration, 43
precipitation, 43
raw material selection, 42
raw materials pretreatment, 42
standardization, 44

L

Limonin extraction
comprehensive extraction process, 52
dichloromethane extraction process, 52
supercritical CO_2 extraction process, 52–53
ultrasonic extraction process, 52
Lipid-soluble carotenoids, 8–9
Low methoxyl pectin
acidification method, 44–45
alkali preparation method, 44
enzymatic method, 45–46
membrane separation, 48
resin/phosphate method, 46–48

M

M-type sprayer, 94–95, 95f

N

Naringin, 7
Neohesperidin, 7–8

P

Pectin, 41–48
high methoxyl pectin. *See* High methoxyl pectin
low methoxyl pectin. *See* Low methoxyl pectin
quality standards, 48, 48t–49t
total pectin, 17–18
water-soluble pectin, 17

2-Phenyl chromone, 4, 5f
Photo-diode array (PAD), 20–21
Pressure spray drying equipment, 93–96
Processed foods
candies
operating points, 80
product quality, 80
technological process, 80
health-promoting citrus peel sauce, 78–79
operating points, 79
quality index, 79
technological process, 78

S

Solid phase microextraction (SPME), 25
Soluble dietary fiber (SDF), 10–11
Spectral analysis, 18–20
Squeezing/precipitation and filtration, 50
S-type sprayer, 96f, 95–96

W

Water-soluble yellow pigments, 8–9

Printed in the United States
By Bookmasters